Lecture Notes in Computer Scien

T0230285

Commenced Publication in 1973
Founding and Former Series Editors:
Gerhard Goos, Juris Hartmanis, and Jan van Leeuwen

Lucas Paletta   John K. Tsotsos
Erich Rome   Glyn Humphreys (Eds.)

Attention and Performance in Computational Vision

Second International Workshop, WAPCV 2004
Prague, Czech Republic, May 15, 2004
Revised Selected Papers

 Springer

Volume Editors

Lucas Paletta
Joanneum Research, Institute of Digital Image Processing
Wastiangasse 6, 8010 Graz, Austria
E-mail: lucas.paletta@joanneum.at

John K. Tsotsos
York University, Department of Computer Science and Center for Vision Research
4700 Keele Street, Ontario, M3J 1P3, Toronto, Canada
E-mail: tsotsos@cs.yorku.ca

Erich Rome
Fraunhofer Insitute for Autonomous Intelligent Systems
Schloss Birlinghoven, 53754 Sankt Augustin, Germany
E-mail: erich.rome@ais.fraunhofer.de

Glyn Humphreys
University of Birmingham, Behavioural Brain Sciences Centre
B15 2TT, Birmingham, UK
E-mail: g.w.humphreys@bham.ac.uk

Library of Congress Control Number: 2004117729

CR Subject Classification (1998): I.4, I.2, I.5, I.3

ISSN 0302-9743
ISBN 3-540-24421-2 Springer Berlin Heidelberg New York

Springer is a part of Springer Science+Business Media

springeronline.com

© Springer-Verlag Berlin Heidelberg 2005
Printed in Germany

Typesetting: Camera-ready by author, data conversion by Scientific Publishing Services, Chennai, India
Printed on acid-free paper   SPIN: 11378754   06/3142   5 4 3 2 1 0

Preface

In recent research on computer vision systems, attention has been playing a crucial role in mediating bottom-up and top-down paths of information processing. In applied research, the development of enabling technologies such as miniaturized mobile sensors, video surveillance systems, and ambient intelligence systems involves the real-time analysis of enormous quantities of data. Knowledge has to be applied about what needs to be attended to, and when, and what to do in a meaningful sequence, in correspondence with visual feedback. Methods on attention and control are mandatory to render computer vision systems more robust.

The 2nd International Workshop on Attention and Performance in Computational Vision (WAPCV 2004) was held in the Czech Technical University of Prague, Czech Republic, as an associated workshop of the 8th European Conference on Computer Vision (ECCV 2004). The goal of this workshop was to provide an interdisciplinary forum to communicate computational models of visual attention from various viewpoints, such as from computer vision, psychology, robotics and neuroscience. The motivation for interdisciplinarity was communication and inspiration beyond the individual community, to focus discussion on computational modelling, to outline relevant objectives for performance comparison, to explore promising application domains, and to discuss these with reference to all related aspects of cognitive vision. The workshop was held as a single-day, single-track event, consisting of high-quality podium and poster presentations. Invited talks were given by John K. Tsotsos about attention and feature binding in biologically motivated computer vision and by Gustavo Deco about the context of attention, memory and reward from the perspective of computational neuroscience.

The interdisciplinary program committee was composed of 21 internationally recognized researchers. We received 20 manuscripts responding to the workshop call for papers; each of the papers was assigned at least 3 double-blind reviews; 16 of the papers were accepted, as they corresponded to the requested quality standards and suited the workshop topic; 10 were attributed to 4 thematic oral sessions, and 6 were appropriate for representation as posters. The low rejection rate was commonly agreed to be due to the high quality of the submitted papers.

WAPCV 2004 was made possible by the support and engagement of the European Research Network for Cognitive Computer Vision Systems (ECVision). We are very thankful to David Vernon (Coordinator of ECVision) and Colette Maloney of the European Commission's IST Program on Cognition for their financial and moral support. We are grateful to Radim Sara, for the perfect local organization of the workshop and the registration management. We also wish to thank Christin Seifert, for doing the difficult task of assembling these proceedings.

October 2004

Lucas Paletta
John K. Tsotsos
Erich Rome
Glyn W. Humphreys

Organization

Organizing Committee

Chair

Lucas Paletta (Joanneum Res., Austria)
John K. Tsotsos (York Univ., Canada)
Erich Rome (Fraunhofer AIS, Germany)
Glyn W. Humphreys (Birmingham, UK)

Program Committee

Minoru Asada (Osaka Univ., Japan)
Gerriet Backer (Krauss SW, Germany)
Marlene Behrmann (CMU, USA)
Leonardo Chelazzi (Univ. Verona, Italy)
James J. Clark (McGill Univ., Canada)
Bruce A. Draper (Univ. Colorado, USA)
Jan-Olof Eklundh (KTH, Sweden)
Robert B. Fisher (Univ. Edinburgh, UK)
Horst-M. Gross (TU Ilmenau, Germany)
Fred Hamker (Univ. Münster, Germany)
John M. Henderson (MSU, USA)

Laurent Itti (USC, USA)
Christof Koch (Caltech, USA)
Bastian Leibe (ETH Zurich, Switzerland)
Michael Lindenbaum (Technion, Israel)
Nikos Paragios (ENPC Paris, France)
Satyajit Rao (Univ. Genoa, Italy)
Ronald A. Rensink (UBC, Canada)
Antonio Torralba (MIT, USA)
Jeremy Wolfe (Harvard Univ., USA)
Hezy Yeshurun (Tel Aviv Univ., Israel)

Sponsoring Institutions

ECVision — European Research Network for Cognitive Computer Vision Systems
Joanneum Research, Austria

Table of Contents

Applications of Attentive Vision

Distributed Control of Attention

Ola Ramström and Henrik I Christensen

KTH, 10044 Stockholm, Sweden
{olar, hic}@nada.kth.se
http://www.nada.kth.se/cvap

Abstract. Detection of objects is in general a computationally demanding task. To simplify the problem it is of interest to focus the attention to a set of regions of interest. Indoor environments often have large homogeneous textured objects, such as walls and furniture. In this paper we present a model which detects large homogeneous regions and uses this information to search for targets that are smaller in size. Homogeneity is detected by a number of different descriptors and a coalition technique is used to achieve robustness. Expectations about size allow for constraint object search. The presented model is evaluated in the context of a table top scenario.

1 Introduction

In everyday life we have the impression to constantly perceive everything in the visual field coherently and in great detail. One would normally expect to notice a gorilla walking across the scene while watching a basketball game. However, we often fail to notice salient events that are not expected [SM01]. Indeed, only a small fraction of the visual properties of a scene is attended and consciously perceived. Tsotsos' complexity analysis [Tso90] concludes that an attentional mechanism, that selects relevant visual features and regions for higher level processes, is required to handle the vast amount of visual information in a scene.

Garner [Gar74] found that similarity between objects is measured differently depending on whether they differ in integral or separable features. From these findings Treisman and Gelade [TG80] developed the "Feature-Integration Theory of Attention", which states that integral features (denoted dimensions) are processed pre-attentively across the visual field. Consequently a target will appear to pop-out if it is unique in one dimension, such as a red target among green distractors. However, in conjunction search, when the target is not uniquely described by any dimension, such as in search for a red vertical target among red horizontal and green vertical distractors, we must inspect each object in turn and hence the search time will be proportional to the number of distractors. The theory furthermore predicts perceptual grouping to be processed pre-attentively across the visual field. In the conjunction search example, a red vertical target among red horizontal and green vertical distractors, the target can appear to pop-out if e.g. all green objects are on the left side and the red are on the right side of a display. The two groups need to be inspected in turn and the red vertical target will pop-out as the only vertical object in the red group.

L. Paletta et al. (Eds.): WAPCV 2004, LNCS 3368, pp. 1–15, 2005.

Wolfe et. al. [WCF89] and others have found many cases where conjunction search is much faster than the Feature-Integration Theory predicts, clearly different search strategies are used depending on the scene-properties. Treisman and Sato revised the theory [TS90] and confirmed the use of multiple strategies. One of these strategies is to inhibit parts of the background; they found that search performance depends on the homogeneity of the background. Apparently, the background context is processed to ease the search for foreground objects. Moreover, many experiments have demonstrated that we detect and implicitly learn unattended background context [KTG92] [DT96] [HK02]. This implicit memory affects our visual search performance but cannot be accessed by our conscious mind. The Inattentional Amnesia Hypothesis [Wol99] explains this as: Although we perceive and process the whole visual field, only attended locations are consciously remembered.

Clearly, the processing of unattended background information plays an important role in object detection.

The Coherence Theory [Ren00] defines the concept of volatile proto-objects that are formed pre-attentively across the visual field. Proto-objects are described as "relatively complex assemblies of fragments that correspond to localized structures in the world"; for example occluded objects are processed to estimated complete objects [ER92]. Attention is needed for proto-objects to become stable and for conscious processes to access its information. When attention is released the proto-objects become volatile again. This implies that there is little short-term memory apart from what is being attended; this is consistent with the Inattentional Amnesia Hypothesis. Recent biological findings [MvE02] confirm pre-attentive processes corresponding to proto-objects in the Coherence Theory.

We propose a model that is inspired by the Coherence Theory in that it models a way for pre-attentive proto-objects to become stable when attended. The assemblies of proto-objects are used as background context and their statistics are used to efficiently search for objects that are defined only by their size.

1.1 Related Work

Most models of visual attention are space based [Mil93] [Wol94] [TSW+95] [IK00]. Some have modeled different aspects of object based visual attention: Li [Li00] has developed a model for pre-attentive segmentation, Sun and Fisher for computing saliency of hierarchical segments and the attentional shift among these [SF03]. However, none of the above models how pre-attentive segments become stable when attended as predicted in the Coherence Theory.

2 Conceptual Model

A model has been developed which searches for target objects of an expected size. The model is designed to be implemented on a distributed system and uses concepts from game theory to minimize the need for inter-process communication.

The strategy is to use knowledge of the environment; e.g. in a living-room we might expect large items with homogeneous surfaces such as table and cupboard. The large

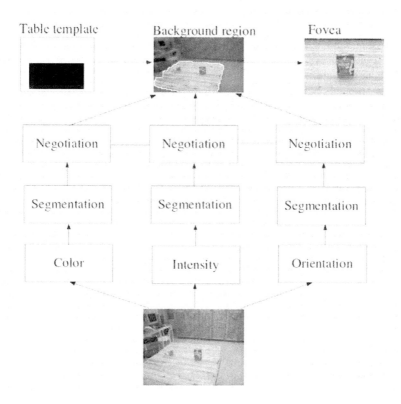

Fig. 1. A raw image is processed by a set of distributed nodes resulting in a set of background regions, which often corresponds to large objects

homogeneous regions provide layout and contextual information of the scene, which can be used to guide the attention.

Figure 1 illustrates the system: A raw image from a camera is decomposed into a set of feature maps at separate nodes, namely color, intensity, and orientation (see section 3). A segmentation algorithm searches for large homogeneous regions locally at each node. The resulting segments are sensitive to variations in the intrinsic parameters and the camera pose, similar to proto objects discussed in [Ren00]. A negotiation scheme forms coalitions of segments, which are more stable than the individual segments, similar to the nexus discussed in [Ren00]. The coalitions of segments are formed by only sharing real valued coalition values across the nodes and the final winning segmentation mask (see section 4). The coalitions are denoted background regions and provide layout information; a spatial template selects interesting background regions. The feature statistics of each interesting background region is computed as local context and saliency can thereafter be computed at each node with respect to the local context. Only a sparse set of the saliency data need to be integrated across the nodes to achieve accurate object detection (see section 5). As reference a center-surround saliency algorithm has been developed (see section 6). The performance of the model is evaluated in section 7.

3   Image Processing

The system processes a raw image into a set of feature maps which are subsequently segmented into a set of homogeneous background regions. The raw image has 300×224 pixels in resolution using the YUV color space and the decomposed feature maps have 75×56 pixels in resolution with different dimensionality. The format of the raw images enable accurate processing of image features and the four times down sampling of the feature maps reduce noise and thus improves extraction of homogeneous regions.

From experimental psychology [ER92] and biology [MvE02] it is clear that the visual cortex performs segmentation preattentively using several separate visual features, Julez denoted such features textons [Jul81]. Treisman [TG80] found that segments cannot be formed by conjunctions of separate features. We will in this model restrict us to three separate feature dimensions: Color, intensity, and orientation. These are suitable to the environment we will use for evaluation. Note that it is not claimed that these are better than any other feature dimensions nor that three separate dimensions is an optimal number of dimensions. However, these feature dimensions allow us to compare the results to [IK00], where similar although not identical features are used.

Feature map identity is denoted $d \in \{color, intensity, orientation\}$ and the feature maps are denoted f^d. In order to make output from the different feature maps comparable when searching for homogeneous regions all feature maps are normalized to have zero mean and unit variance.

4   Background Regions

The processing of feature maps is distributed over a set of processing nodes. Distributed computing increases the processing power if the communication across nodes is limited. It is of interest to enable distributed control where the nodes processes a majority of information locally and only integrates a small subset of the full data set.

This is enabled using background regions, which are created using a game theoretic negotiation scheme.

Knowing the context of a homogeneous background region we can efficiently search for target objects, e.g. knowledge of the appearance of a tablecloth can efficiently guide search for a cup on a table.

Having a rough estimate of the pose of a table relative to the camera, we need a mechanism to find background regions, which might represent large homogeneous objects and are stable with respect small variations in camera pose.

The mean-shift algorithm [CM99] is a fairly well established segmentation algorithm based on local similarities. It has two intrinsic parameters: spatial scale s and feature range r. Small changes in these two intrinsic parameters or in the camera pose can result in different segmentation results.

However, by using redundant mean-shift segments, corresponding to different parameters, we can increase the stability. We will in this section define a negotiation scheme to form coalitions of similar mean-shift segments. Such coalitions are used to extract a segmentation mask which is more stable than the ingoing members.

4.1 Clustering of Redundant Mean-Shift Segments

The mean-shift segmentation depend on two intrinsic parameters (r, s). Different values of these parameters might result in different segmentation result. Since they relate to distances in the spatial and feature domain, such variations are related to variations in pose and illumination. To increase the stability we compute the mean-shift segmentation varying $r \in M_r$ and $s \in M_s$. We will in this work restrict $M_r = \{2, 3, 4\}$ and $M_s = \{0.15, 0.22, 0.33\}$. Furthermore, since we are interested in background regions which are much larger than the expected target size, we select only the $N = 4$ largest segments and discard all segments smaller than 400 pixels (10% of the feature map size) for each selection of $(r, s) \in M_r \times M_s$.

The mean-shift segmentation algorithm is processed locally at each node for each $(r, s) \in M_r \times M_s$. Let P^d represent the resulting set of mean-shift segments at node d; hence the size of $|P^d| \le N|M_s \times M_r|$. Each mean-shift segment $p_i \in P^d$ is associated with a segmentation mask S_i^d and a histogram of the feature values inside the segmentation mask h_i^d. The similarity between two mean-shift segments p_i and p_j is defined as the normalized intersection of their segmentation masks and histograms:

$$Sim(i, j, d) = \frac{S_i^d \cap S_j^d}{|S_i^d| + |S_j^d|} \cdot \frac{h_i^d \cap h_j^d}{|h_i^d| + |h_j^d|} \tag{1}$$

The second factor is fairly standard in histogram matching, the first factor borrows the same normalization technique and gives a penalty when they differ spatially in size, location, and shape.

To find the optimal background regions we need to evaluate all possible cluster combinations of mean-shift segments, which has exponential complexity $O(2^{|P^d|})$. This complexity is reduced using a modified version of the coalition formation process proposed by [SK98], which only have square complexity with respect to the number of mean-shift segments $O(|P^d|^2)$.

4.2 Negotiation

Coalitions of mean-shift segments are formed by an iterative negotiation process. At iteration $t = 0$ each mean-shift segment $p_i \in P^d$ broadcast its description, (S_i^d, h_i^d), and selects a set of possible coalition members $C_i^d(0) \subseteq P^d$, including all other mean-shift segments with a spatial similarity larger than th:

$$C_i^d(0) = \{\forall p_j \in P^d | Sim(i, j, d) > th\} \tag{2}$$

We define the value of a coalition $C_i^d(t)$ at iteration t as:

$$V_i^d(t) = \sum_{j \in C_i^d(t)} Sim(i, j, d) \tag{3}$$

mean-shift segments are valued relative to their contribution, hence the value of p_j in coalition $C_i^d(t)$ is:

$$V_i^d(j, t) = Sim(i, j, d) \tag{4}$$

Thus, each segment forms a coalition including similar segments at the same node. From this set of coalitions, background regions are iteratively extracted by a distributed negotiation scheme. In each iteration of the negotiation the strongest coalition across all nodes is chosen and used to form a background region and to inhibit segments in succeeding negotiation iterations.

In more detail, stable coalitions are formed when each mean-shift segment p_i at each node d iteratively perform the following:

1. Compute and announce $V_i^d(t) = \sum_{j \in C_i^d(t)} V_i^d(j, t)$ to all other segments at all nodes.
2. Choose the highest among all announced coalition values, $V_{max}(t)$.
3. If no other coalition at any node has a stronger coalition value, $V_i^d(t) = V_{max}(t)$, then compute the weighted segmentation mask $W^d = q \sum_{j \in C_i^d} V_i^d(j, t) S_j^d$; where $q \in \mathbf{R}^1$ is a constant which normalize W^d to have maximal value one. Remove all $p_i \in C_i^d$ from further negotiation.
4. Update $V_i^d(j, t+1) = (1 - 2\frac{S_i^d \cap S_{max}^d}{|S_i^d| + |S_{max}^d|}) V_i^d(j, t)$;
5. Start over from 1.

At each iteration a weighted segmentation mask, W^d, is computed and $|C_i^d|$ is decreased for at least one mean-shift segment. The process will be repeated until all C_i^d are empty or for a fixed number of iterations.

Note that the set of nodes only compares values of maximal coalitions, all other computation of image data is processed locally at each node. The weighted segmentation masks resulting from a raw image illustrated in figure 2.

4.3   Segmentation Mask

A segmentation mask can be extracted by thresholding each weighted segmentation mask W^d. However, we do not have any analytic way to extract such threshold value. Instead we calculate a Gaussian-mixture model (GMM) of the complement region of W^d in the associated feature map f^d, using 5 Gaussian models.

The resulting probability maps are suitable competition to W^d. We compute the E_{ij} maps, which is the probability of pixel i to belong to the Gaussian model j, for each $j = 1, 2, 3, 4, 5$. Furthermore, we denote the Gaussian model at W^d with $j = 0$, hence E_{i0} is the probability map for pixel i to belong to the Gaussian model at W^d. Finally, the probability for pixel i to belong to background region C_{max}^d is the joint probability $W_i^d E_{i0}$.

The segmentation mask, S^d, associated with C_{max}^d is defined as the pixels i where $W_i^d E_{i0} > E_{ij}$ for all $j = 1, 2, 3, 4, 5$.

The segmentation mask from a raw image is illustrated in figure 3.

4.4   Region Completion

[ER92] demonstrated in a series of experiments that the visual cortex preattentively completes homogeneous regions to object hypothesis. Inspired by this result we perform region completion of the segmentation masks S^d. The object completion is not as advanced as the completion process demonstrated by [ER92], however it enables

Fig. 2. Raw image and weighted summation mask for first, second, and third background region

Fig. 3. Raw image and segmentation mask for first, second, and third background region

detection of objects within homogeneous regions. One obvious completion process is to fill in holes. Furthermore, objects e.g. at the border of a table often pop-out from the table leaving a notch on the border of the segmentation mask. Regions corresponding to artificial large indoor-objects, in the set of evaluation scenes, are often square or have some vertical or horizontal straight lines. Following this discussion, we define the region completion as filling in gaps where a vertical or horizontal straight line can be attached to the original segment. Note that a notch in the corner will not be completed by this process as predicted by [ER92] .

Moreover, we do not want to detect other overlapping background regions as salient. Therefore we restrict object completion to regions not occupied by other original segmentation masks.

Figure 4 illustrates a completed segmentation mask. We observe that the region is more compound.

This process is solely based on intuition from the [ER92] experiments and can obviously be improved. However, the completion process enables accurate objects detection and is sufficient at this point in the development process.

Fig. 4. Raw image and completed segmentation mask for first, second, and third background region

5   Target Objects

All background regions that are spatially similar to a template are attended, e.g. all background regions in the lower part of the visual field. The feature statistics of each background regions is extracted and used to search for outliers which are likely to be a target objects.

This process involves sending a sparse set of conspicuous data to all other nodes. After this integration of sparse data, salient locations can be attended by foveated vision, and the target hypothesis can be figure-ground segmented.

5.1   Background Region of Interest

The task is to find objects on a table which are expected to occupy a large fraction of the lower half of the scene. We use a template to represent this knowledge. A background region which overlaps more than 25% with the template it is considered a background region of interest. Figure 5 illustrates the template used here.

Fig. 5. Template for background region

5.2   Foreground Region of Interest

To find foreground regions of interest (ROI) we calculate the feature statistics of each background region of interest. The weighted summation mask W^d of each such background region is used to calculate (m^d, Σ^d) at each node $d \in \{color, intensity, orientation\}$:

$$m^d = \frac{1}{|W^d|} \sum_{x=0}^{X} \sum_{y=0}^{Y} f^d(x, y) W^d \tag{5}$$

$$\Sigma^d = \frac{1}{|W^d|} \sum_{x=0}^{X} \sum_{y=0}^{Y} (f^d(x, y) - m^d)(f^d(x, y) - m^d)^T W^d \tag{6}$$

where W^d is the sum of all pixels in W^d.

We calculate the set of pixels p_S^d which can be excluded from background region S^d with confidence γ with respect to a Normal distribution (m^d, Σ^d). If this set is larger that q, the confidence value is increased and a new set p_S^d is computed. This process is repeated until the size $|p_S^d| < q$.

In the current implementation we have chosen the set $\gamma \in 0.5, 0.6, 0.7, 0.8, 0.9, 1$ and $q = 0.25|S|$ without further investigation.

The sparse set of conspicuous pixels p_S^d is distributed to all other nodes. At each node an integrated saliency map is constructed from the sum of all p_S^d. The integrated saliency map in convolved with a Gaussian kernel with a standard deviation equal to the expected target size. Each peak of the saliency map, which is larger than one, is extracted as ROI. Hence, we only consider regions, of expected target size, that at least one node has found conspicuous.

Each peak, which is selected in the saliency map, is attended with a foveated camera with sixteen times higher resolution, right-most image in figure 6.

Fig. 6. Top row: Raw image, background region, and interest points at background region Bottom row: foveated view at interest points

6   Center-Surround Saliency

A well-established attention model is the center-surround saliency model developed by [IK00]. The source code is available at `http://ilab.usc.edu/toolkit/`. However, to suite our choices of feature map definitions, an own implementation has been developed based on [IK00]. Our implementation uses fewer scales and integral feature maps (described in section 3), and is hence not equally good as the original. However, it has similar properties as the original and is used here as a comparison of typical behavior. It will be denoted CS-search.

It should be pointed out that CS-search does not have the same focus on distributed processing as the proposed model; at the final step complete pixel maps are integrated.

6.1   CS-Search

Using the integral images we define center-surround saliency as the Euclidean distance between the mean feature vector inside a center rectangle (rc) and the mean feature vector inside a surrounding larger rectangle (rs). Let (rc, rs) denote a center rectangle with width rc and a surrounding rectangle with width rs, as illustrated in figure 7.

Six different center-surround saliency maps $CS^d_{(rc,rs)}$ are computed at each node with:

$$(rc, rs) \in \{(20, 50); (20, 60); (30, 60); (30, 70); (40, 70); (40, 80)\}$$

and

$$d \in \{color, intensity, orientation\}$$

Fig. 7. Center-surround rectangles (rc,rs)

The 18 center-surround saliency maps (six maps at three nodes) are integrated using a weighted summation

$$CS_{tot} = \sum_{\forall (rc,rs)} \sum_{\forall d} w^d_{(rc,rs)} CS^d_{(rc,rs)}$$

The weights, $w^d_{(rc,rs)}$, are defined as the square of the difference between the maximal peak value and the mean peak value at each center-surround saliency map. Peaks are defined as local maxima with respect to a four-connected neighborhood. The weights act as lateral normalization: Center-surround maps with many equally strong peaks are attenuated whereas maps a few salient peaks are amplified. See [IK00] for more details.

If we search for cups, or simply want to inspect the scene for objects of target size, we do not want to stop after the first cup or item is found. Rather, we want to generate a scan path for attentive scrutinizing. We simulate interest points for such a scan path by the ten strongest CS-search peaks within 50% of the strongest peak.

In figure 8, right most image, we see that the box is detected as a salient point but not the cup, given 10 points within 50% of the strongest. It is not surprising that the clutter to the left in the image attract more attention than the cup on the table.

Fig. 8. Raw image, center-surround saliency map, and saliency peaks (50%)

We address these drawbacks of CS-search with the proposed model, as illustrated in figure 6: From a raw image we extract a background region which is likely to correspond to the table and search for target hypotheses using its context. We observe that using the context of the tabletop, only the box and the cup are considered salient.

7   Result

There is no single correct ground truth in segmentation and attention; both vary with scale and task, among many other factors. To avoid such issues the system has been evaluated in terms its ability to detect targets on a table. The detection performance will depend on its ability to extract background regions.

For the evaluation five different scene set-ups has been used and in each scene the camera pose and the targets on the table has been varied in eight different configurations. Figure 9 lists a subset of 40 resulting evaluation raw images. The scenes were chosen so the proposed model performs well but not perfect. Quite naturally the proposed model does not perform well in cluttered scenes where no homogeneous regions are found. The selection of raw images provides a sense of what the proposed algorithm considers a non-cluttered scene.

We observe that in seven of the eight configurations of each scene there are two objects on the table, in one configuration the table is empty. Thus, there is a total of $2 \times 7 \times 5 = 70$ objects to be detected on 40 different appearances of the table.

In section 7.1 an experiment is presented which illustrates the power of foveated vision and the need for an accurate peripheral attention mechanism. In section 7.3 the proposed model is evaluated as peripheral attention mechanism.

Fig. 9. Raw images used in the experiments

Fig. 10. Box and cup used as targets

7.1 SIFT Recognition

A recognition experiment has been done in cooperation with Fredrik Fuesjö [RF04]. SIFT features described in [Low99] was used for recognition on both the peripheral and foveated view. The peripheral view correspond to the set of raw images in figure 9 and foveated view correspond to the set of views foveated by the attentional mechanism.

The texture of the cup and box (illustrated in figure 10) enables accurate detection by the SIFT features and were therefore selected as targets. Both targets are present in 15 of the 40 raw images. All redundant saccades and other objects are candidates for false positive responses.

The SIFT recognition algorithm results in a set of matches between the observed image and a memory. By applying a threshold on number necessary of matches it is possible to determine if the image correspond to the memory or not. The results are presented as ROC curves in figure 11. We observe that both the box and cup is close to chance performance in the peripheral view and close to perfect in foveated view.

It can be concluded that recognition in the peripheral view is just above chance but nearly perfect by foveated vision.

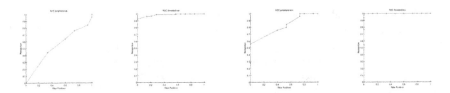

Fig. 11. ROC curves. Left to right: cup target for peripheral view, cup target for foveated view, box target for peripheral, and box target for foveated view

7.2   CS-Search

In CS-search we do not attempt to extract any background context other that a local surrounding, see chapter 6 for details. Since the information about large homogeneous regions is omitted, a salient table may attract many interest points around its border and between targets. Moreover, since saliency is computed on a global scale a target, which is salient with respect to the table, may not be salient enough compared to the rest of the scene.

Using each object on the table in figure 9 as targets, the CS-search did find 62 out of the 70 possible targets. A total of 350 interest points was generated, which correspond to about nine points per image and target detection at 18% of the interest point.

It should be repeated that the CS-search only extracts the ten most salient points within 50% of the saliency of the strongest peak. A short evaluation indicated that the chosen thresholds are fairly good; to increase the detection rate many more irrelevant interest points was generated and hence decreasing the 18% detection rate.

7.3   Background Region Extraction

In the proposed model background regions are extracted which provides information of the layout of the scene. A table template, illustrated in figure 5, is used to take advantage of this information: the table is said to be extracted when at least one extracted background region spatially intersects 25% of the table template .

Moreover, the feature distribution of the background regions can be used as contextual information in target detection. The target objects are salient enough to be detected if the table is well extracted as a background region. However, if the background region is larger than the table the feature variance is increased and the objects can escape detection. On the other hand, if the background does not extract the whole table, the objects might not be included in the region and therefore not detected.

As a reference, each feature dimension and each parameter selection which was used redundantly in the proposed model, has been evaluated separately and the result is presented in the table below. Each row represent one selection of (s, r)-Mean-Shift parameters and each column represent one selection of feature dimension ("combined" is a feature vector $\in \mathbf{R}^6$ with color, intensity, and orientation concatenated). The results are presented as

(#tables extracted, #targets detected); where the maximum is (40,70).

(s,r)	color	intensity	orientation	combined
(2,0.1)	(5,5)	(23,38)	(8,6)	(3,3)
(2,0.1)	(17,20)	(26,37)	(18,22)	(6,6)
(2,0.1)	(17,22)	(27,39)	(19,25)	(8,11)
(3,0.15)	(13,15)	(28,42)	(14,17)	(9,11)
(3,0.15)	(18,27)	(25,40)	(21,26)	(16,13)
(3,0.15)	(17,27)	(19,25)	(30,34)	(17,19)
(4,0.225)	(17,20)	(24,36)	(21,22)	(12,12)
(4,0.225)	(15,24)	(19,26)	(28,33)	(17,23)
(4,0.225)	(15,23)	(16,14)	(32,42)	(17,25)

We can conclude that no single selection of Mean-Shift parameter and feature dimension is appropriate for all scenes and scene variations in this evaluation. The orientation node with $(s, r) = (4, 0.225)$ was best by extracting the table in 32 instances and detecting the target in 42 instances.

Moreover, from inspection it was observed that the table in each raw image always was well segmented by at least one selection of feature dimension and Mean-Shift parameter. In more detail, the raw images on row 1-2 in figure 9 was well segmented by the color dimension, row 3-4 by the orientation node, and row 5-10 by the intensity dimension.

It can be concluded that with appropriate choice of feature dimension and Mean-Shift parameters the table can always be well segmented in each of the 40 raw images, but no single such choice is appropriate for all raw images.

The performance of the proposed model is presented below for a set of different similarity thresholds, th, in the coalition formation (see chapter 4).

Th	#tables extracted	#targets detected	#interest points
0.4	70	66	137
0.5	70	66	137
0.6	70	67	131
0.7	70	65	125
0.8	70	66	123
0.9	69	64	132

We observe that the proposed model always extract the table and 65 to 67 targets, except for $th = 0.9$. The number of interest points range from 123 to 137 and hence the detection rate from 48% to 54%.

It can be concluded that although the extraction of the table is not always good enough to detect the targets, the coalition scheme clearly improves the table extraction. The number of detected targets is better than what can be achieved with any of the other evaluated methods and the detection rate is superior compared to CS-search.

It can be concluded that extraction of background regions adds layout and context information which prunes the number of interest points and increases target saliency.

8   Summary

An attentional mechanism on a peripheral view, which guides a foveated view, does indeed improve recognition. Since saccades take time it is of interest to prune the number of irrelevant saccades.

A comparison to a CS-search mechanism demonstrated that the extracted layout information clearly decreased the number of irrelevant interest points. Moreover, it was demonstrated that no single selection of feature dimension and mean-shift algorithm parameters could perform as well as the proposed model; the clustering among redundant segments clearly improved the quality of contextual information at the background regions.

Furthermore, the model enables distributed processing. Common bottlenecks in distribute processing are the need for sharing data and central control. The proposed model resolves these issues by applying game theoretical concepts where only a sparse subset of the data is shared in a distributed coalition formation scheme.

In conclusion, the proposed model enables distribute processing by extracting information of the layout to reduce the number of irrelevant interest points and the context of homogeneous regions to increase target saliency and hence increase the target detection rate.

References

[CM99]   D. Comaniciu and P. Meer. Mean shift analysis and applications. *Proc. Seventh Int'l Conf. Computer Vision*, pages 1197–1203, September 1999.

[DT96]   B. DeSchepper and A. Treisman. Visual memory for novel shapes: Implicit coding without attention. *Journal of Experimental Psychology: Learning, Memory, and Cognition*, 22:27–47, 1996.

[ER92]   J.T. Enns and R.A. Rensink. An object completion process in early vision. *Investigative Ophthalmology & Visual Science*, 33(1263), 1992.

[Gar74]   W.R. Garner. *The processing of information and structure*. Hillsdale NJ: Erlbaum, 1974.

[HK02]   J-S. Hyun and M-S. Kim. Implicit learning of background context in visual search. Technical Report 7, Department of Psychology,, Yonsei University, Seoul, Korea, 2002.

[IK00]   L. Itti and K. Koch. A saliency-based search mechanism for overt and covert shifts of visual attention. *Vision Research*, 40:1489–1506, 2000.

[Jul81]   B. Julez. Textons, the elements of texture perception and their interactions. *Nature*, 290:91–97, 1981.

[KTG92]   D. Kahneman, A. Treisman, and B.J. Gibbs. The reviewing of object files. *Cognitive Psychology*, 24(2):175–219, 1992.

[Li00]   Z. Li. Pre-attentive segmentation in the primary visual cortex. *Spatial Vision*, 13(1):25–50, 2000.

[Low99]   D.G. Lowe. Object recognition from local scale-invariant features. In *Proc. of the International Conference on Computer Vision ICCV, Corfu*, pages 1150–1157, 1999.

[Mil93]   R. Milanese. *Detecting Salient Regions in an Image:From Biological Evidence to Computer Implementation*. PhD thesis, University of Geneva, 1993.

[MvE02]   D.S. Marcus and D.C. van Essen. Scene segmentation and attention in primate cortical areas v1 and v2. *J Neurophysiol*, 88(5):2648–58, November 2002.

[Ren00]   R.A. Rensink. The dynamic representation of scenes. *Visual Cognition*, 7(1):17–42, 2000.

[RF04]   O. Ramström and F. Fuesjö. Recognition with and without attention. CogVis (IST-2000-29375) Deliverable, Enclosure 5 to DR.1.5, June 2004.

[SF03]   Y. Sun and R. Fisher. Object-based visual attention for computer vision. *Artificial Intelligence*, 146:77–123, 2003.

[SK98]   O. Shehory and S. Kraus. Methods for task allocation via agent coalition formation. *Artificial Intelligence*, 101(1–2):165–200, 1998.

[SM01]   D.J. Simons and S.R. Mitroff. The role of expectations in change detection and attentional capture. In M. Jenkin and L. Harris, editors, *Vision & Attention*, chapter 10, pages 189–208. Springe Verlag, 2001.

[TG80]   A. Treisman and G. Gelade. A feature-integration theory of attention. *Cognitive Psychology*, 12:97–136, 1980.

[TS90]   A. Treisman and S. Sato. Conjunction search revisited. *Journal of Experimental Psychology:Human Perception and Performance*, 16(3):459–478, 1990.

[Tso90]   J.K. Tsotsos. Analyzing vision at the complexity level. *Behavioral and Brain Science*, 13:423–469, 1990.

[TSW+95] J.K. Tsotsos, .M. Sean, W.Y.K. Wai, Y. Lai, N. Davis, and F. Nuflo. Modelling visual attention via selective tuning. *Artificial Intelligence*, 78:507–545, 1995.

[WCF89]   J. Wolfe, K. Cave, and S. Franzel. Guided search: An alternative to the feature integration model for visual search. *Journal of Experimental Psychology: Human Perception and Performance*, 15(3):419–433, 1989.

[Wol94]   J.M. Wolfe. Guided search 2.0: A revised model of visual search. *Psychonomic Bulletin & Review I*, 2:202–238, 1994.

[Wol99]   J.M. Wolfe. Inattentional amnesia. In Coltheart, editor, *Fleeting Memories*, pages 71–94. Cambridge, MA: MIT Press, 1999.

Inherent Limitations of Visual Search and the Role of Inner-Scene Similarity

Tamar Avraham and Michael Lindenbaum

Computer Science Department,
Technion, Haifa 32000, Israel
tammya,mic@cs.technion.ac.il

Abstract. This work focuses on inner-scene objects similarity as an information source for directing attention and for speeding-up visual search performed by artificial vision systems. A scalar measure (similar to Kolmogorov's ϵ-covering of metric spaces) is suggested for quantifying how much a visual search task can benefit from this source of information. The measure provided is algorithm independent, providing an inherent measure for tasks' difficulty, and can be also used as a predictor for search performance. We show that this measure is a lower bound on all search algorithms' performance and provide a simple algorithm that this measure bounds its performance from above. Since calculating a metric cover is NP-hard, we use both a heuristic and a 2-approximation algorithm for estimating it, and test the validity of our theorem on some experimental search tasks. This work can be considered as an attempt to quantify Duncan and Humphreys' *similarity theory* [5].

1 Introduction

Visual search is required in situations where a person or a machine views a scene with the goal of finding one or more familiar entities. The highly effective visual search (or more generally, attention) mechanisms in the human visual system were extensively studied from Psychophysics and Physiology points of view. Yarbus [26] found that the eyes rest much longer on some elements of an image, while other elements may receive little or no attention. Neisser [13] suggested that the visual processing is divided into pre-attentive and attentive stages. The first consists of parallel processes that simultaneously operate on large portions of the visual field, and form the units to which attention may then be directed. The second stage consists of limited-capacity processes that focus on a smaller portion of the visual field. Triesman and Gelade's *feature integration theory* [21], aiming to explain the variability of search behavior, suggest that some visual properties (such as color, orientation, spatial frequency, movement) are extracted in parallel across the visual field. Under this hypothesis, if a target is unique in such a feature, it *pops-out* and gets immediate attentional priority. Otherwise, it is detected only after a longer serial search. While several aspects of the Feature Integration Theory were criticized, the theory was dominant in visual search research and much work was carried out, e.g. to understand how feature integration occurs (some examples are [10],[25],[23]). Duncan and Humphreys rejected the dichotomy to parallel vs. serial search and proposed an alternative theory

L. Paletta et al. (Eds.): WAPCV 2004, LNCS 3368, pp. 16–28, 2005.

based on similarity [5]. According to their theory, two types of similarities are involved in a visual search task: between the objects in the scene, and between these objects and the prior knowledge about the possible targets. They suggest that structural units which are similar to the hypothesized target get higher priorities. Moreover, when a scene contains several similar structural units, the search is easier since there is no need to treat every unit individually. Thus, if all non-targets are homogeneous, they may be rejected together resulting in a fast (pop-out like) detection process, while if they are more heterogeneous the search is slower.

Several search mechanisms were implemented, usually in the context of HVS (Human Visual System) studies (e.g. [10], [23], [25], [7]). Other implementations focused on computer vision applications (e.g. [9], [19], [20]), and sometimes used other sources of knowledge to direct visual search. For example, one approach is to search first for a different object, easier to detect, which is likely to appear close to the sought for target ([16], [24]). In [2] we suggested an algorithm based on a stochastic approach for visual search based mainly on inner-scene similarity. In [3] we have suggested a way to extend the same algorithm to use also bottom-up and top-down information, when available.

Relatively little was done to quantitatively characterize the inherent difficulty of search tasks. Tsotsos [22] considers the complexity of visual search and proves for example that spatial boundedness of the target is essential to make the search tractable. In [24], the efficiency of indirect search is analyzed, suggesting a way to decide whether prior information on spatial relationship is helping. A few studies suggest simple measures for quantifying the saliency of a target aiming to explain biology phenomena. For instance, [4] suggests that search is easy if the target is linearly separable from the distractors in color space, and [17] suggests the Mahalanobis distance as a measure for saliency of a target in a scene with moving objects.

In this work we aim to quantitatively characterize the inherent difficulty of search tasks for computer algorithms. As observed in [5], similarity between objects of the same identity can accelerate the search. Assuming that visual similarity is a cue for similar identity, similar objects can be rejected together. The results described in this paper indicate for which scene this assumption is true and for which this approach can be misleading. For most natural scenes the inner-scene-similarity is very informative and can play a major role in improving search performance, especially when top-down information is not available. We characterize the difficulty of the search task using a metric-space cover (similar to Kolmogorov's ϵ-covering [11]) and derive bounds on the performance of all search algorithms. We also propose a simple algorithm that provably meets these bounds.

Interestingly, while we did not aim at modelling the HVS attention system, it turns out that it shares many of its properties and in particular is similar to Duncan and Humphreys's model [5]. As so, our work can be considered as a quantification of their observations.

Paper outline: The context for visual search and some basic intuitive assumptions are described in section 2. Section 3 includes an analytic analysis of search algorithms performance bounds providing measures for search tasks difficulty, a simple deterministic search algorithm, and some intuitive illustration of the results. Section 4 describes some experiments, demonstrating the validity of the bounds and the performance of the suggested algorithm.

2   Framework

2.1   The Context - A Visual Search Task Involving Candidate Selection and Classification

The task of looking for object/s of certain identity in a visual scene is often divided into two subtasks: One is to *select* sub-images which serve as *candidates*. The other, the *object recognition* task, is to decide whether a candidate is a sought for object or not.

The candidate selection task can be performed by a segmentation process or even by a simple division of the image into small rectangles. The candidates may be of different size, bounded or unbounded [22], and can also overlap. The object recognizer is commonly computationally expensive, as the possible appearances of the object may vary due to changes in shape, color, pose, illumination and other imaging conditions. The recognizer may need to recognize a category of objects (and not a specific model), which usually makes it even more complex.

The object recognition process gets the candidates, one by one, after some ordering. This ordering is the attentional mechanism on which we focus here.

2.2   Sources of Information for Directing the Search

Several information sources enabling more efficient search are possible:

Bottom-Up Saliency of Candidates - In modelling HVS attention, it is often claimed that a saliency measure, quantifying how every candidate is different from the other candidates in the scene, is calculated ([21][10][9]). Often, saliency is important in directing attention to objects in a scene. Nevertheless, it can be sometimes misleading and may also not be applicable when, say, the scene contains of multiple resembling targets.

Top-Down Approach - When prior knowledge on the targets is available, the candidates may be ranked by their degree of consistency, or similarity with the target description ([25], [8]). In many cases, however, it is hard to characterize the objects of interest visually in a way which is effective for discriminations and inexpensive to evaluate.

Mutual Similarity of Candidates - Usually, a higher inner-scene visual similarity implies a higher likelihood for similar (or equal) identity ([5]). Under this assumption, after the identity of one (or few) candidates is already known, it can effect the likelihood of the remaining candidates to have the same/different identity.

In this work we focus on (the less studied item -) mutual similarity between candidates, and aim to provide measures for predicting search time, and for quantifying the difficulty of search tasks.

To quantify similarity, we embed the candidates as points in a metric space with distances reflecting dissimilarities. In section 4 we demonstrate some methods for expressing similarity by comparing texture and color of objects in the scene.

2.3   Algorithms Framework

Abstractly, we look at all search algorithms as entities that choose some ordering of the candidates. This order, indicating the search path, can be static or dynamic. Most systems modelling top-down or bottom-up attention propose a precalculated static order as a saliency map. In [2] we proposed a dynamic 'ordering algorithm' suggesting to dynamically change this order each time new information is collected. The new information comes from the results of recognition of already attended candidates. In section 3 we suggest a simplified algorithm (FLNN-Farthest Labelled Nearest Neighbor) that is also dynamic in this sense. We present measures that suffice both as upper-bounds on FLNN's performance and as lower-bounds on the performance of all search algorithms under this framework.

2.4   The Cost of Visual Search

We shall be interested in predicting the search performance. We assume that only the costs associated with calling some *recognition oracle* are substantial. Note that all the other information, including similarity calculation, may be calculated efficiently and even in parallel. Our measures bound the number of calls to the oracle required until finding a target. Our results hold both for scenes including one target and for scenes with multiple targets. (In the second case it will provide the cost for finding the first target.)

3   Bounds on Visual Search Performance

The simplest search situation is when there is one target in the scene, and all other candidates are very different from the target and mutually-similar. The hardest case is when all mutual similarities of candidates are alike and thus un-informative. Such situations happen when all candidates are very similar or when all candidates are different from each other. An intermediate difficulty search task is, for example, when the candidates can be divided into a few groups, each group members having strong mutual similarity. The search task characteristic, developed in this section, meaningfully quantifies the search task difficulty.

3.1   Notations

We consider an abstract description of a search task as a pair (X, l), where $X = \{x_1, x_2, \ldots, x_n\}$ is a set of partial descriptions associated with the set of candidates, and $l : X \rightarrow \{T, D\}$ is a function assigning identity labels to the candidates. $l(x_i) = T$ if the candidate x_i is a target, and $l(x_i) = D$ if x_i is a non-target (or a distractor).

An attention, or search algorithm, A, is provided with the set X, but not with the labels l. It requires $\text{cost}_1(A, X, l)$ calls to the recognizer oracle, until the first target is found.

We refer to the set of partial descriptions $X = \{x_1, x_2, \ldots, x_n\}$ as points in a metric space $(S, d), d : S \times S \rightarrow \mathbb{R}^+$ being the metric distance function. The partial description can be, for example, a feature vector, and the distance may be the Euclidian metric.

3.2   A Difficulty Measure Combining Targets Isolation and Candidates Scattering

The proposed measure for task difficulty combines two main factors:

1. The distance between target candidates and non-target candidates.
2. The distribution of the candidates in the feature space.

Intuitively, the search is easier when the targets are more distant from non-targets. But, if the non-targets are also different between themselves, the search becomes hard again.

A useful quantification for expressing a distribution of points in a metric space (and thus expressing how scattered the candidates are) uses the notation of metric cover.

Definition 1. *Let $X \subseteq S$ be a set of points in a metric space (S, d). Let 2^S be a set of all possible subsets of S. $C \subset 2^S$ is 'a cover' of X if $\forall x \in X \exists C \in \mathcal{C}$ s.t. $x \cap C \neq \emptyset$.*

Definition 2. *$\mathcal{C} \subset 2^S$ is a 'd_0-cover' of a set X if $\mathcal{C}$ is a cover of X and if $\forall C \in \mathcal{C}$ $diameter(C) < d_0$, where $diameter(C)$ is $\max_{c_1, c_2 \in C} d(c1, c2)$.*

Definition 3. *A 'minimum-d_0-cover' is a d_0-cover with a minimal number of elements. We shall use the notation $\mathcal{C}_{d_0}(X)$ for some particular minimum-d_0-cover, and denote its size by $c_{d_0}(X)$.*

If, for example, X is a set of feature vectors in a Euclidian space, $c_{d_0}(X)$ is the minimum number of m-spheres with diameter d_0 required to cover all points in X. The above definitions follow Kolmogorov's ϵ-covering [11].

Definition 4. *Given a search task (X, l), let the 'max-min-target-distance', denoted d_T, be the largest distance of a target to its nearest non-target neighbor.*

The main result of this section is:

Theorem 1. *Given a search task (X, l), so that d_T is its max-min-target-distance, the difficulty of finding the first target depends on the minimum-d_T-cover size, $c_{d_T}(X)$, in the sense that:*

1. *Any search algorithm in the worst case needs to query the oracle for at least $c_{d_T}(X)$ candidates before finding a target.*
2. *There is an algorithm that needs no more than $c_{d_T}(X)$ queries for finding the first target.*

The proof to the two parts of the theorem follows after some intuition on this result: Consider the cases described in the beginning of section 3. When one target differs dramatically from all non-targets, which are similar, $c_{d_T}(X)$ is 2. (All non-targets can be covered together by one covering element, and the target alone is covered by a second). In a case where all n candidates are scattered and all mutual distances are $\geq d_T$, $c_{d_T}(X)$ is n. When there are k groups of candidates so that all inner-group distances $< d_T$ and all outer-group distances $\geq d_T$, $c_{d_T}(X)$ is k.

The above examples are simple extreme cases in which the difficulty is intuitively revealed. $c_{d_T}(X)$ also provides a measure for intermediate situations which are harder to analyze. For instance, the candidates may be scattered, but a target can be significantly far away from any of them, providing a relatively small $c_{d_T}(X)$. On the other hand, the candidates may be divided into a small number of groups, but since the targets and non-targets are mixed together in one of those groups (or more), $c_{d_T}(X)$ will be relatively big, characterizing the search difficulty correctly. Note also that, naturally, the cover size, and the implied search difficulty, depend on the partial description (features), which may be chosen depending on the application.

Proof to Theorem 1.1 The claim in theorem 1.1 defines a *lower bound* on all search algorithms performance. Let us first phrase it more accurately:

$$\forall A \; \exists X \in \mathcal{X}_{d_0,c}, l \in \mathcal{L}_{d_0} \; \mathrm{cost}_1(A, X, l) \geq c$$

where $\mathcal{X}_{d_0,c}$ denotes all sets of candidates for which the minimum-d_0-cover-size is c, and $\mathcal{L}_{d_0}$ denotes all labels assignments for which max-min-target-distance $d_T \geq d_0$. The claim in words: For any search algorithm, we can provide a search task for which it will find a target after at least c queries.

Proof: Choose c points in the metric space, so that all the inner-point distance is more than d_0. Choose the n candidates to de divided equally between these locations, providing the input set of candidates X.

Until a search algorithm A finds the first target, it receives only *no* answers from the recognition oracle. Therefore, given an input candidate set X, and a specific algorithm A, the sequence of attended candidates may be simulated under the assumption that the oracle returns only *no* answers. Let π be this resulting ordering of X.

Choose an assignment of labels l that assigns T only to the group of candidates located in the point that its first appearance in π is last. A will query the oracle at least c times before finding a target.   ∎

Note that no specific metric is considered in the above claim and proof.

Proof to Theorem 1.2 In theorem 1.2 we claim that there exists a search algorithm for which $c_{d_0}(X)$ is an *upper bound* on its performance. We suggest the following simple algorithm, which suffices for the proof:

FLNN- Farthest Labelled Nearest Neighbor: Given a set of candidates $X = \{x_1, \ldots, x_n\}$, choose randomly the first candidate, query the oracle and label this candidate. Repeat iteratively, until a target is detected: for each unlabelled candidate x_i, compute the distance dL_i to the nearest labelled neighbor. Choose the candidate x_i for which dL_i is maximum. Query the oracle to get its label.

Let us now again phrase the claim more accurately:

$$\forall X \in \mathcal{X}_{d_0,c}, l \in \mathcal{L}_{d_0} \; \mathrm{cost}_1(\mathrm{FLNN}, X, l) \leq c$$

or in words: if there is a target which is more than d_0 far from all the distractors ($d_T \geq d_0$), and all the candidates can be covered by c sets with diameter at most d_0 , FLNN finds the first target after at most c queries.

Proof: Take an arbitrary minimum-d_0-cover of X, $\mathcal{C}_{d_0}(X)$. Let x_i be a target so that $d(x_i, x_j) \geq d_0$ for every distractor x_j. (Since $d_T \geq d_0$ such an x_i exists.) Let C be a covering element($C \in \mathcal{C}_{d_0}(X)$) so that $x_i \in C$. Note that all candidates in C are targets. Excluding C, there are $(c-1)$ other covering elements in $\mathcal{C}_{d_0}(X)$ with diameter $< d_0$. Since C contains a candidate that its distance from all distractors $\geq d_0$, FLNN will not query two distractor-candidates in one covering element (that their distance $< d_0$), before it queries at least one candidate in C. Therefore, a target will be located after at most c queries. (it is possible that a target that is not in C will be found earlier, and then the algorithm stops even before).   ∎

Discussion. Note that although the performance of FLNN depends on d_T and the minimum cover size, the algorithm itself is independent from this information. Also note, that for the case of a single target, a more simple algorithm that queries the oracle by the descending order of their distance to their nearest neighbor, also provides the same upper bound. When there is a possibility for multiple targets, such an algorithm might perform very bad (it is likely that the targets are feature-space-close) and FLNN is required.

The above measure can be used for predicting search performance. Say we have prior knowledge that for the family of search tasks we are dealing with d_T always exceeds some d_0. Then, for any specific task X, $c_{d_0}(X)$ can be used as a predictor for its search time.

The problem of finding the minimum cover is NP-hard. Gonzalez [6] proposes a 2-approximation algorithm for the problem of clustering a data set minimizing the maximum inner-cluster distance, and proves it is the best approximation possible if $P \neq NP$. Gonzalez's algorithm promises a result that is an upper bound on the cover size and always smaller than twice the cover size, and therefore provides a lower and an upper bound on the real cover size. In the experiments we use a heuristic algorithm that provides tighter upper bounds, and the FLNN's worst cases results serve as a tighter lower bound.

Note that FLNN suffice for proving the complexity tightness, but suffers from several drawbacks: It relates only to the nearest neighbor, which makes it non-robust: a single attended distractor close to an undetected target, reduces the priority of this target and slows the search. Moreover, it does not extend naturally to finding more than one target, and to incorporating bottom-up and top-down information when available. For a more practical algorithm, addressing these problems, we direct the reader to [3].

Illustration of Above Results. We demonstrate the latest result using the 'insects' example in figure 1. (For more realistic examples see section 4.) The image contains 7 different insects, which are roughly segmented. For simplicity of the demonstration, the search uses only the objects area as a partial description (feature). See the distribution of the objects in the one-dimensional feature space in figure 1c. Taking 'ladybug' as the target, the *minimum-d_T-cover* is of size 3. Therefore, using FLNN algorithm, the ladybug is detected after examining at most 3 objects. Note that if the image had some more butterflies and some more small creatures, like ants, $c_{d_T}(X)$ for the target 'ladybug' would not have been changed. Also note, that if there where some more ladybugs in the image, using FLNN algorithm we could still promise that one of the ladybugs would have been found after the same number of queries promised before. In these cases the search difficulty wouldn't have grown with additional candidates. This comes to illustrate that when using a dynamic attention ordering as we suggest, we can be less picky in the candidate selection process, and leave more candidates without paying a large price during the search.

3.3   Bounds Relying on Less Information

In this section we provide some less tighter bounds that rely on weaker information. In the above sections we assumed that for some d_0 the minimum-d_0-cover-size ($c_{d_0}(X)$) of a search task X is known, or is possible to compute. In addition, we are pre-informed

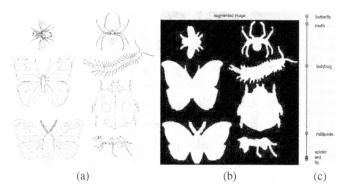

(a)   (b)   (c)

Fig. 1. 'insects' : Image for demonstrating the minimum-d_T-cover-size bound. (a)The original 'insects' image. (b)The 'insects' image after a rough segmentation. (c)Insects in one dimensional feature space. feature = object area

Table 1. Bounds on search performance relying on weaker information. The knowledge in each row is added to the knowledge in the previous row. Euclidean metric is considered, but similar results for other metrics are easy to derive. While the minimum-d_0-cover size is tighter, number of non-empty spheres (last bound) is easier to compute

Knowledge	Bound
No knowledge	$|X|$
A lower bound d_0 on d_T	$|\{x_i \in X \mid d_i > d_0\}|$
The metric space is bounded: $X \subset [0,1]^m$	$\min(\lceil \frac{\sqrt{m}}{d_0} \rceil^m, |X|)$
The metric space is sparse	Number of non-empty covering spheres

that there exists a target in the scene that its (feature-space) distance from all distractors is at least d_0. Sometimes such information is not available. Then, one may use looser bounds, which are briefly summarized in table 1. The bounds are valid and tight in the sense suggested in Theorem 1. For full proofs, see [1]. When no information is available, the only bound possible to suggest is the trivial, which is the number of candidates. When only a lower bound d_0 on d_T is known, we suggest to count the number of candidates with distance to their nearest-neighbor bigger than d_0 as a lower bound for performance. When the metric space is bounded by for instance the hypercube $[0,1]^m$, the number of spheres with diameter d_0 required to cover this hypercube, which is $\lceil \frac{\sqrt{m}}{d_0} \rceil^m$, can act as a lower bound for all algorithms and as an upper bound for FLNN. Tightening the last measure, from all $\lceil \frac{\sqrt{m}}{d_0} \rceil^m$ covering spheres we can count only the ones occupied by candidates. Minimum-d_0-cover-size is always tighter, but the number of non-empty spheres is easier to compute.

4   Experiments

Our first set of experiments uses the COIL-100 database [14], using the 100 database objects in one pose each (36x36 gray scaled images). We think of these images as

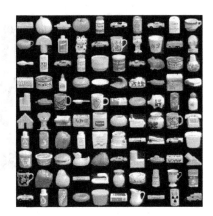

Fig. 2. Columbia100 - COIL-100 database images

candidates extracted from some larger image. See figure 2 for a mosaic image denoted *columbia100* .

The extracted features are gaussian derivatives, based on the filters described in [15]. We use two first order gaussian derivatives (at 0 and 90 degrees), three second order gaussian derivatives (at 0, 60 and 120 degrees), and four third order gaussian derivatives (at 0, 45, 90 and 135 degrees). Each of them at 5 scales, giving a total of 45 filters, resulting in feature vectors of length 45 per candidate. Euclidian metric measures the feature vectors distance.

First, we choose *cups* as the targets category (10 out of the 100 candidates are cups). We ran the FLNN 100 times, starting from a different candidate each run. The maximal number of queries before finding the first target was 18 (min = 1, average $\approx$ 9).

For this same search task we compute an approximation of the minimum-d_T-cover-size. Since computing the optimal result is NP-complete, we use the following heuristic procedure which we found to give a close upper bound on the real value: We start with computing d_T (definition 4). Then, iteratively, we choose a candidate for which the number of d_T-close candidates is minimal. We check if this set of candidates can be covered by one covering element (in many cases they do). If not, we remove items from this set until its diameter doesn't exceed d_T. The resulting set is removed from the set of whole candidates, and the procedure is repeated until all candidates are covered. The number of covering elements used in this process is the output of the algorithm. Note that this algorithm gives a cover size which is valid, but may not be minimal. We have also implemented the 2-approximation algorithm suggested by Gonzalez [6]. The results of the algorithm described above gave a better result for all cases.

For the case of the cups search task, the heuristic algorithm found a cover of size 24. The 2-approximation algorithm's result was 42, providing a possible range of 21 to 42. From those results together we are left with a lower and upper bound for the real minimal cover size of 21 and 24. As expected, FLNN's worst performance (18) is below.

Still using the columbia100 image, we repeat the above set of experiments for the case where the targets category is *toy cars*, and for the case where targets category is *toy animals*. The results are summarized in table 2 (first three rows) and in figure 4 .

Fig. 3. The *elephants* and *parasols* images taken from the Berkeley hand segmented database and the segmentations we used in our experiments. (colored images)

Table 2. Experiments results for FLNN and minimal cover size. The numbers next to the search tasks names are the number of targets and the number of overall candidates in that task. The real value of minimal cover size is bounded from below by 'FLNN worst' and the half of '2-Approx. cover size', and bounded from above by 'Heuristic cover size' and '2-Approx. cover size'

Search task	# of cand.	# of targets	FLNN worst	FLNN mean	Heuristic cover size	2-Approx. cover size	Real cover size
cups	100	10	18	8.97	24	42	21-24
cars	100	10	73	33.02	79	88	73-79
toy animals	100	7	22	9.06	25	42	22-25
elephants	24	4	9	5.67	9	11	9
parasols	30	6	6	3.17	8	13	7-8

Comparing the three above cases, we can see that for the *cars* case we get the worst performance, while for the two other cases results are similar. In the *cars* case, the targets are very similar to each other, which should ease the search. Yet, finding the first car is hard since there are some distractors which are very similar to the cars. (Here d_T is small.) In the *cups* case most cups are similar and they are all quite different from distractors, causing the task of finding the first cup to be of reasonable difficulty. The results of the *toy animals* case are similar, but not due to the same reasons. The different toy animals have no resemblance between themselves, but one of them (a duck) is very different from all candidates and is easy to find.

The next two experiments use different input images, a different method for candidates selection, and different type of feature vectors. The two images, the *elephants* image and the *parasols* image, are taken from the Berkeley hand segmented Database [12]. See figure 3 for the two images and the hand segmentations we used. Segments under a certain size are ignored, leaving us with 24 candidates in the *elephants* image and 30 candidates in the *parasols* image. The targets are the segments containing elephants and parasols respectively. (For some colored images, including the parasols image, a simple automatic color-segmentation process provided segments which we were also able to use as input candidates to the search algorithms. The results in this case were similar to the ones we present here for the hand-segmented candidates.)

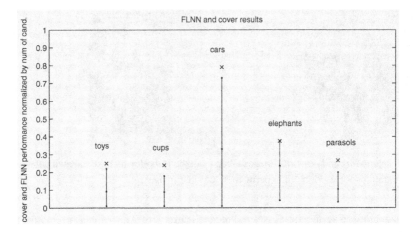

Fig. 4. Graphic representation of FLNN and cover size experiments results. The results in table 2 are normalized by the number of candidates to enable comparison between difficulty of different search tasks. The vertical lines describe the results of all FLNN runs, the dots being the best, mean and worst results. The x describe the heuristic calculated cover size

For those colored images we use color histograms as feature vectors. In each segment (candidate), we extract the values of $\frac{b}{r+g+b}$ and $\frac{r}{r+g+b}$ from each pixel, where r, g, and b are values from the RGB representation of a colored image. Each of these two dimensions are divided into 8 bins. This results a 2D histogram with 64 bins, i.e - a feature vector of length 64. Again, we use Euclidean metric for distance measure. (We tried also other histogram comparison methods, such as the ones suggested in [18], but the results where similar.) The results for those two search tasks also appear in table 2 and figure 4.

Note: Table 2 deals with estimating the absolute number of objects that are attended before a target is found. Figure 4 demonstrates the amount of objects attended during the search, relatively to the number of objects in the scene (by dividing the values in table 2 by the number of candidates of the corresponding search task).

5 Discussion and Future Work

In this paper we considered the usage of inner-scene similarity for visual search and focus on its inherent limitations. The cover size which quantifies the search difficulty combines both the targets/non-targets similarity and the non-targets/non-targets similarity, suggested in [5] as the qualitative characteristics determining human visual search performance. The quantitative measures provided here allow not only to measure difficulty but also to predict the search time. Not surprisingly, our results easily recognize the two extreme situations of 'pop-out' and 'sequential' searches, while locating each search task in a point on a continuous axis between these two poles.

While the study of feature selection is usually in the context of object recognition, here we show that it is very important for visual search as well.

We see the results suggested in this paper as a first step for quantifying visual search performance. Here we have suggested a worst case analysis and studied only the role of inner-scene similarity. We hope to provide in the future a statistic analysis and measures taking into account also bottom-up and top-down available information. Also, as suggested in the WAPCV04 (Second International Workshop on Attention and Performance in Computer Vision, Prague, May 2004) after this paper was presented, we are now checking whether there is a connection between the measures suggested here and the performance of biology visual search.

References

1. T. Avraham and M. Lindenbaum. Dynamic visual search using inner scene similarity - algorithms and bounds. *Technion. Computer Science Departement. Technical Report CIS-2003-02*, May 2003.
2. T. Avraham and M. Lindenbaum. A probabilistic estimation approach for dynamic visual search. *In Proceedings of WAPCV03 - First International Workshop on Attention and Performance in Computer Vision*, pages 1–8, April 2003.
3. T. Avraham and M. Lindenbaum. Dynamic visual search using inner scene similarity - algorithms and inherent limitations. *In Proceedings of ECCV04 - The 8th European Conference on Computer Vision*, LNCS 3022:pages 58–70, May 2004.
4. B. Bauer, P. Jolicoeur, and W. B. Cowan. Visual search for colour targets that are or are not linearly separable from distractors. *Vision Research*, 36:1439–1465, 1996.
5. J. Duncan and G.W. Humphreys. Visual search and stimulus similarity. *Psychological Review*, 96:433–458, 1989.
6. T.F. Gonzalez. Clustering to minimize the maximum intercluster distance. *Theoretical Computer Science*, 38(2-3):293–306, June 1985.
7. G.W. Humphreys and H.J. Muller. Search via recursive rejection (serr): A connectionist model of visual search. *Cognitive Psychology*, 25:43–110, 1993.
8. L. Itti. Models of bottom-up and top-down visual attention. *Thesis*, January 2000.
9. L. Itti, C. Koch, and E. Niebur. A model of saliency-based visual attention for rapid scene analysis. *PAMI*, 20(11):1254–1259, November 1998.
10. C. Koch and S. Ullman. Shifts in selective visual attention: towards the underlying neural vircuity. *Human Neurobiology*, 4:219–227, 1985.
11. A.N. Kolmogorov and V.M. Tikhomirov. $epsilon$-entropy and $epsilon$-capacity of sets in functional spaces. *AMS Translations. Series 2*, 17:277–364, 1961.
12. D. Martin, C. Fowlkes, D. Tal, and J. Malik. A database of human segmented natural images and its application to evaluating segmentation algorithms and measuring ecological statistics. In *Proc. 8th Int'l Conf. Computer Vision*, volume 2, pages 416–423, July 2001.
13. U. Neisser. *Cognitive Psychology*. Appleton-Century-Crofts, New York, 1967.
14. S. Nene, S. Nayar, and H. Murase. Columbia object image library (coil-100). *Technical Report CUCS-006-96, Department of Computer Science, Columbia University*, February 1996.
15. R.P.N. Rao and D.H. Ballard. An active vision architecture based on iconic representations. *Artificial Intelligence*, 78(1–2):461–505, 1995.
16. R.D. Rimey and C.M. Brown. Control of selective perception using bayes nets and decision theory. *International Journal of Computer Vision*, 12:173–207, 1994.
17. R. Rosenholtz. A simple saliency model predicts a number of motion popout phenomena. *Vision Research*, 39:3157–3163, 1999.
18. M.J. Swain and D.H. Ballard. Color indexing. *International Journal of Computer Vision*, 7:11–32, 1991.

19. H. Tagare, K. Toyama, and J.G. Wang. A maximum-likelihood strategy for directing attention during visual search. *IEEE PAMI*, 23(5):490–500, 2001.

20. A. Torralba and P. Sinha. Statistical context priming for object detection. In *Proceedings of the Eighth International Conference On Computer Vision*, pages 763–770, 2001.

21. A. Treisman and G.Gelade. A feature integration theory of attention. *Cognitive Psychology*, 12:97–136, 1980.

22. J.K. Tsotsos. On the relative complexity of active versus passive visual search. *IJCV*, 7(2):127–141, 1992.

23. J.K. Tsotsos, S.M. Culhane, W.Y.K. Wai, Y. Lai, N. Davis, and F.J. Nuflo. Modeling visual attention via selective tuning. *Artificial intelligence*, 78(1-2):507–545, 1995.

24. L.E. Wixson and D.H. Ballard. Using intermediate objects to improve the efficiency of visual-search. *IJCV*, 12(2-3):209–230, April 1994.

25. J.M. Wolfe. Guided search 2.0: A revised model of visual search. *Psychonomic Bulletin and Review*, 1(2):202–238, 1994.

26. A.L. Yarbus. *Eye Movements and Vision*. Plenum Press, New York, 1967.

Attentive Object Detection Using an Information Theoretic Saliency Measure

Gerald Fritz[1], Christin Seifert[1], Lucas Paletta[1], and Horst Bischof[2]

[1] JOANNEUM RESEARCH Forschungsgesellschaft mbH,
Institute of Digital Image Processing,
Wastiangasse 6, A-8010 Graz, Austria
{gerald.fritz, christin.seifert, lucas.paletta}@joanneum.at
[2] Graz University of Technology,
Institute for Computer Graphics and Vision,
Inffeldgasse 16/II, A-8010 Graz, Austria
bischof@icg.tu-graz.ac.at

Abstract. A major goal of selective attention is to focus processing on relevant information to enable rapid and robust task performance. For the example of attentive visual object recognition, we investigate here the impact of top-down information on multi-stage processing, instead of integrating generic visual feature extraction into object specific interpretation. We discriminate between generic and specific task based filters that select task relevant information of different scope and specificity within a processing chain. Attention is applied by tuned early features to selectively respond to generic task related visual features, i.e., to information that is in general locally relevant for any kind of object search. The mapping from appearances to discriminative regions is then modeled using decision trees to accelerate processing. The focus of attention on discriminative patterns enables efficient recognition of specific objects, by means of a sparse object representation that enables selective, task relevant, and rapid object specific responses. In the experiments the performance in object recognition from single appearance patterns dramatically increased considering only discriminative patterns, and evaluation of complete image analysis under various degrees of partial occlusion and image noise resulted in highly robust recognition, even in the presence of severe occlusion and noise effects. In addition, we present performance evaluation on our public available reference object database (TSG-20).

1   Introduction

A major issue in cognitive computer vision is to investigate the scheduling of information between bottom-up and top-down paths of visual information processing. A viewpoint commonly found in the visual attention literature is to extract regions of interest from generic interest operators, and proceed with task specific image processing thereafter on this selected information. This methodology is widely accepted as valid model for both machine and human vision. In contrast, recent investigations in neurobiology suggest that visual attention modulates neural activity throughout all levels of visual cortex [2], including primary visual cortex, and task based models for machine attention have been

L. Paletta et al. (Eds.): WAPCV 2004, LNCS 3368, pp. 29–41, 2005.

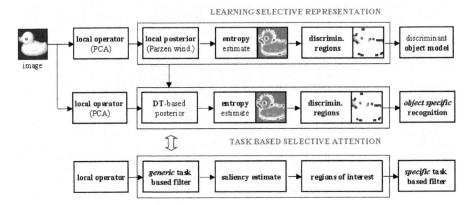

Fig. 1. Concept of the multi-stage task based attention model. (Bottom line) Most general, locally extracted information can be processed by several task based filters that range from generic to more specific characteristics (left to right). (Center line) In case of attentive object recognition, salient regions can be extracted from a filter that responds to object information in general (left - using a decision tree based approach), while object specific information can be derived in a second stage (right). (Top line) The discriminative filters are learned during an extensive off-line processing stage

reported with superior performance [20]. A thorough analysis on task specific interest operators and how they contribute to visual attention has not been given so far.

We investigate in this work visual attention on the task of object recognition, using an information theoretic saliency measure for the determination of discriminative regions of interest. Previous work on the exploitation of local operator responses for attentive object search developed several viewpoints on how to advance from local to global information. In object recognition from local information, [22] applied standard interest operators (Förstner, Harris) with the aim to determine localizable object parts for further analysis. [22] selected class related patterns from related fixed spatial configurations of recognizable operator responses. To avoid dependency on scale selection, [10, 12, 14] introduced interest point detectors that derive scale invariance from local scale saliency. These operators proved to further improve recognition from local photometric patterns, such as in [4]. While concern has been taken specifically with respect to issues of scale invariance [9, 11, 10], wide baseline stereo matching performance [12, 14], or unsupervised learning of object categories [22, 1, 4], the application of interest point operators has not yet been investigated in depth about the information content they provide with respect to a visual task, such as, object discrimination. While [8] determine saliency from a class-independent local measure of sample density, proving superior performance to generic interest operators, [21] went further by outlining a selection process on informative image fragments as features for classification, confirming performance improvements using specific in contrast to generic features.

The key contribution of this work to visual attention is to investigate information theoretic saliency measures with respect to object search and recognition, and to settle a process chain from generic to specific task based feature extraction with the goal of cascaded attention. We firstly determine a detailed quantitative analysis of the discriminative power of local appearances, and, secondly, exploit discriminative object regions to build up an efficient local appearance representation and recognition methodology as an example for generic to specific task based attention (Fig. 1). In contrast to the use of classifiers that determine discriminative features (e.g., [19]) for recognition, our approach intends to make the actual local information content explicit for further processing, such as, constructing the object model (Sec. 2) or determining discriminative regions for recognition and detection (Sec. 3). In this way we intend to extend the work of [8, 21], firstly, by providing a precise estimate of local information content, secondly, by evaluating the robustness of the approach with respect to various degrees of noise and partial occlusion, and further, by tentatively setting recognition from local information content in a multi-stage frame on attentive processing.

We propose in a first stage to localize discriminative regions in the object views from the Shannon entropy of a locally estimated posterior distribution (Sec. 2.1). In a second stage, we consequently derive object models in feature subspace from discriminative local patterns (Sec. 2.2). Object recognition is then exclusively applied to test patterns with associated low entropy. Identification is achieved by majority voting on a histogram over local target attributions (Sec. 3). Rapid object recognition is supported by decision tree bases focus of attention on discriminative regions of interest (Sec. 4). The recognition method is evaluated on images degraded with Gaussian noise and different degrees of partial occlusions using the COIL database (Sec. 5), while the attention mechanism is tested on detecting objects of interest from the COIL database within a cluttered background.

2   Entropy-Based Object Models

The proposed object model consists of projections of those local appearances that provide rich information about an object identity, i.e., *reference imagettes*[1] mapped into a subspace of the corresponding image matrix. Local regions in the object views that are both discriminative and robustly indicate the correct object label provide the reference imagettes for the object representation.

Note that discriminative regions correspond here to regions in the image that contain information with respect to the task of object recognition. Feature extraction is here applied from principal component analysis and followed by an estimate of the local posterior to derive the local information content. In contrast to comparable work on the extraction of discriminative basis libraries (e.g., [18]) for an efficient construction of feature subspaces, we specifically exploit the local information content value to derive efficient thresholds for pattern rejection, with the purpose to determine both representation and region of interest.

2.1   Local Distributions in Subspace

We use a principal component analysis (PCA, [13]) calculated on local image windows of size $w \times w$ to form the basis for our local low dimensional representation. PCA maps

---

[1] Imagettes denote subimages of an object view [3].

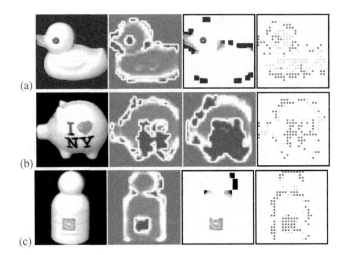

Fig. 2. Sample COIL-20 objects (a) o_1, (b) o_{13}, (c) o_{16} at view 0; each with - left to right - (i) original frame, (ii) entropy image (from 9x9 pixel imagettes; entropy coded color palette from low=blue to high=red), (iii) local appearances with $\Theta \leq 0.5$ in a,c (entropy image from 15x15 pixel imagettes in b), and (iv) accuracy-coded images (accuracy blue=true, white=false)

imagettes $\mathbf{x}_i$ to a low dimensional vector $\mathbf{g}_i = \mathbf{E}\mathbf{x}_i$ by matrix $\mathbf{E}$ consisting of few most relevant eigenvectors of the covariance matrix about the imagette sample distribution $\mathbf{X} = \{\mathbf{x}_1, \ldots, \mathbf{x}_i, \ldots, \mathbf{x}_N\}$, N is the total number of imagette samples.

In order to get the information content of a sample $\mathbf{g}_i$ in eigenspace with respect to object identification, we need to estimate the entropy $H(O|\mathbf{g}_i)$ of the posterior distribution $P(o_k|\mathbf{g}_i)$, $k = 1 \ldots \Omega$, Ω is the number of instantiations of the object class variable O. The Shannon entropy denotes

$$H(O|\mathbf{g}_i) \equiv -\sum_k P(o_k|\mathbf{g}_i) \log P(o_k|\mathbf{g}_i). \tag{1}$$

We approximate the posteriors at $\mathbf{g}_i$ using only samples $\mathbf{g}_j$ inside a Parzen window [16] of a local neighborhood ϵ, $\|\mathbf{g}_i - \mathbf{g}_j\| \leq \epsilon$, $j = 1 \ldots J$. We weight the contributions of specific samples $\mathbf{g}_{j,k}$ - labelled by object o_k - that should increase the posterior estimate $P(o_k|\mathbf{g}_i)$ by a Gaussian kernel function value $\mathcal{N}(\mu, \sigma)$ in order to favour samples with smaller distance to observation $\mathbf{g}_i$, with $\mu = \mathbf{g}_i$ and $\sigma = \epsilon/2$. The estimate about the Shannon entropy $\hat{H}(O|\mathbf{g}_i)$ provides then a measure of ambiguity in terms of characterizing the information content with respect to object identification within a single local observation $\mathbf{g}_i$.

2.2 Discriminative Object Regions

It is obvious that the size of the local ϵ-neighborhood will impact the distribution and thereby the recognition accuracy (Fig. 3, Sec. 5), highly depending on the topology of object related manifolds in subspace. One can construct an entropy-coded image of an object view from a mapping of local appearances $\mathbf{x}_i$ to corresponding entropy estimates

(Fig. 2 for characteristic objects of the COIL-20 database, Sec. 5). The individual images (from left to right) depict the original image, the color coded entropy image (from 9×9 pixel imagettes), corresponding imagettes with $\hat{H}(O|\mathbf{g}_i(\mathbf{x}_i)) \leq 0.5$, and associated images coding recognition accuracy (blue=correct, white=false). From these images it becomes obvious that regions containing specific texture and brightness contrasts provide highly discriminative information for recognition.

From discriminative regions we proceed to *entropy thresholded* object *representations*. The entropy coded images provide evidence that segmentation into discriminative regions and consequently exclusive usage of the associated reference points in eigenspace would provide sparse instead of extensive object representations [3], in terms of storing only imagette information that is *relevant for classification* purposes. Object representations from local photometric patterns have been constructed either from extensive storage of all subspace (reference) points for k-nearest neighbor matching [3], or from selected, cluster specific prototype points [22, 4] that necessarily convey uncertainty. In contrast, the proposed object model includes only *selected* reference points for nearest neighbor classification, storing exclusively those $\mathbf{g}_i$ with

$$\hat{H}(O|\mathbf{g}_i) \leq \Theta. \tag{2}$$

A specific choice on the threshold Θ consequently determines both storage requirements and recognition accuracy (Sec. 5). To speed up the matching we use efficient memory indexing of nearest neighbor candidates described by the adaptive *K-d* tree method [5].

3   Object Recognition from Local Information

The proposed recognition process is characterised by an entropy driven selection of image regions for classification, and a voting operation, as follows,

1. **Mapping** of imagette patterns into subspace (Sec. 2.1).
2. **Probabilistic interpretation** to determine local information content (Eq. 1).
3. **Rejection** of imagettes contributing to ambiguous information (Sec. 2.2).
4. **Nearest neighbor analysis** of selected imagettes within ϵ-environment.
5. **Hierarchical voting** for object identifications integrating local voting results.

Each imagette pattern from a test image that is mapped to an eigenspace feature point is analysed for its entropy $\hat{H}(O|\mathbf{g}_i)$ with respect to object identification. In case this imagette would convey ambiguous information, its contribution to a global recognition decision would become negligible, therefore it is removed from further consideration. Actually, practice confirms the assumption that it is difficult to achieve a globally accurate object identification when multiple ambiguous imagettes 'wash out' any useful evidence on a correct object hypothesis [15]. The entropy threshold Θ for rejecting ambiguous test points in eigenspace is easily determined from the corresponding threshold applied to get the sparse model of reference points by Eq. 2. Selected points are then evaluated on whether they lie within the ϵ-distance of any model reference point.

Majority voting on the complete set of local decisions from feature responses has already been proven to represent a powerful method for object recognition [3, 6]. However, detection tasks require the localization of object regions within cluttered background or other objects of interest. We therefore propose a *hierarchical voting* schema that would integrate voting results from a local neighborhood into more global regions of interest. Fig. 8 illustrates all relevant processing stages of the recognition and voting process, for the example of two object images impacted by severe occlusion effects. Note that the entropy images contain discriminative regions (blue) that partially are not represented in the object voting map, which is due the fact that some feature samples project outside the entropy selected object features. E.g., the local voting for object o_{14} results in five tiles labelled for the true object and one outlier, which will be rejected by the global majority voting integration step.

4   Attentive Object Detection from Local Information

For the purpose of *rapid* object recognition and detection, we need a mapping of low computational complexity to perform a *focus of attention* on regions of interest (ROIs). These ROIs would then be fed into the recognition module (Sec. 3) for detailed analysis. Actually, this segmentation function would work in terms of a point of interest operator (POI) but be tuned from a discriminative objective function, i.e., entropy based. In order to keep complexity of the POI low, we do not use a neural network approach or a universal function estimator. In contrast, we provide rapid entropy estimates from a decision tree classifier, assuming appropriate quantization of the entropy values into class labels. One expects from this tree to provide this estimate from a few attribute queries which would fundamentally decrease computation time per image for ROI computation.

Estimation of Entropy Values.   For a rapid estimation of local entropy quantities, each imagette projection is fed into the decision tree which maps eigenfeatures $\mathbf{g}_i$ into entropy estimates $\hat{H}$, $\mathbf{g}_i \mapsto \hat{H}(\Omega|\mathbf{g}_i)$. The C4.5 algorithm [17] builds a decision tree using the standard top-down induction of decision trees approach, recursively partitioning the data into smaller subsets, based on the value of an attribute. At each step in the construction of the decision tree, C4.5 selects the attribute that maximizes the information gain ratio. The induced decision tree is pruned using pessimistic error estimation [17].

Rapid Extraction of ROIs.   The extraction of ROIs in the image is performed in 2 stages. First, the decision tree based entropy estimator provides a rapid estimate of local information content. Only eigenfeatures $\mathbf{g}_i$ with an associated entropy below a predefined threshold $\hat{H}(O|\mathbf{g}_i) < H_\Theta$ are considered for recognition (Sec. 3). These selected discriminative eigenfeatures are then processed by nearest neighbor analysis with respect to the object models and by majority voting according to the process described in Sec. 3.

5   Experimental Results

In order to perform a thorough analysis of the object recognition performance we applied the described methodology to images of the COIL-20 database [13].

Eigenspace Representation.   Experiments were applied on 72 (test: 36) views of 20 objects of the COIL-20 database, for a total of 1440 (720) gray-scaled images with normalized size of 128×128 pixels. Analysis was performed with various imagette sizes: 9×9 (selected for further experiments), 15×15, and 21×21 pixels, resulting in larger discriminative regions (normalised per entropy maximum; e.g., in Fig. 2b) for larger imagette sizes. However, recognition errors due to partial occlusion positively correlate with imagette size. Imagettes were sampled by a step size of 5 pixels, giving a total of 603950 (301920) imagettes for training (test), excluding black background imagettes. Imagettes were projected into a 20-dimensional eigenspace.

Local Information Content. For a local probabilistic interpretation of test imagettes, we searched for an appropriate threshold ϵ to determine the training samples in a local neighborhood (Sec. 2.1) that will be considered for the posterior estimation. Fig. 3 shows recognition rates (using a MAP classifier and images degraded with 10% Gaussian noise, operated on all *single* imagettes) with various values for neighborhood ϵ and entropy threshold Θ. This diagram shows that imagettes with high entropy $\Theta > 2$ dramatically decrease the recognition rate. $\epsilon = 0.1$ was selected for best performance while not taking too many imagettes into account.

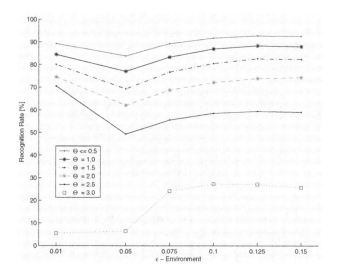

Fig. 3. Recognition performance using MAP classification on samples of a neighborhood ϵ. Rejecting imagettes with entropy $\hat{H}(O|\mathbf{g}_i) > \Theta, \Theta = 2.0$, may dramatically increase accuracy of overall object recognition

Discriminative Regions from Decision Trees. The decision tree was trained using eigenfeatures $\mathbf{g}_i$ of 50% of all extracted imagettes from the COIL-20 database and associated entropy estimates, determined from the Parzen window approach (Sec. 2.1). Entropy values were linearly mapped into equally spaced N_H intervals $(k - 1)H_{max}/N_H$, $kH_{max}/N_H]$ for $k = 1..N_H$. In our experiments N_H was set to 5 (Tab. 1). The error on the training set was determined 2.4%, the error on the test set 13.0%. This suffices to map eigenfeatures very efficiently to corresponding entropy intervals (classes). Fig. 4

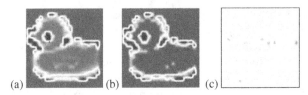

Fig. 4. Estimation of entropy from decision trees. (a) Entropy estimation from Parzen windows I_P (Sec. 2.1, color code see Fig. 2), (b) entropy estimation from decision tree I_T (Sec.4), (c) difference image $I_D = I_P - I_T$ (grayvalue coded for $[0, H_{max}]$ from white to black)

illustrates a typical decision tree based estimation of local entropy values. It demonstrates that discriminative regions can be represented highly accurately (Fig. 4b) from sequences of only ≈ 25 attribute queries. A difference image (Fig. 4c) reflects the negligible errors and confirms that discriminative regions can be both reliably and rapidly estimated from C4.5 decision trees.

Table 1. Confusion map of the C4.5 decision tree based entropy estimation. The individual entropy intervals - denoted by classes $c_1 \cdots c_5$ - partitioning $[0, H_{max}]$ into equally large intervals (Sec. 5) are well mapped by the decision tree to corresponding output classes, providing an accurate estimation of the local entropy values (Fig. 4)

maps $\longmapsto$	c1	c2	c3	c4	c5
c1	38146	5546	2981	1425	218
c2	1820	44418	1382	624	72
c3	1495	2094	41989	2548	190
c4	1036	1136	3725	40775	1644
c5	166	192	400	2709	44849

Partial Occlusion and Gaussian Noise. For a thorough performance analysis in case of image corruption, we applied partial occlusion of $0 - 90\%$ and Gaussian noise to pixel brightness. For determining the occlusion area, we selected random center positions of the black occluding squared windows within the object regions, and computed the window size related to the total number of pixels attributed to a given object. This prevents from preferring specific object regions for occlusions, and assures that the object is actually occluded according to a given occlusion rate. Fig. 5 depicts sample entropy coded color images corrupted by various degrees of occlusion. The associated histograms on imagette based object label attribution illustrate that majority voting mostly provides a both accurate and robust decision on the object identity.

Indoor Object Recognition Performance. The experiments on recognition rates from occlusion and noise demonstrate the superior performance of the entropy critical method as well as the associated majority voting classifier. Fig. 6 demonstrates best performance for an interpretation of a complete image that has been treated by small Gaussian noise $= 10\%$. However, for detection issues we have to confine to local regions such as those

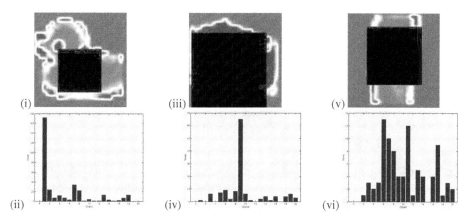

Fig. 5. Samples of (top down) 40%, 80% and 60% occlusion on entropy coded images and associated histogram on imagette based object classes. Majority voting provided (top down) correct/correct/incorr. identification

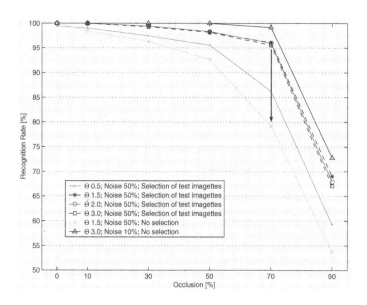

Fig. 6. Resulting recognition performance for up to 90% occlusion rates and Gaussian noise of 50% (10%)

segmented by entropy thresholding ($\Theta < \Theta_{max}$). Several models of entropy thresholding demonstrate the robustness of the system performance with respect to Gaussian noise = 50% and for varying degrees of occlusions. Note that with an entropy critical selection of 30% ($\Theta = 1.5$) out of all possible test imagettes an accuracy of > 95% is achieved despite a 70% occlusion rate (blue). Considering instead *all* test imagettes for recognition (no selection), the performance would drop by more than 15% (green;

Fig. 7. Sample results of attentive object detection (Sec. 5. (a) Original frame with COIL-20 objects superimposed. (b) Detected object locations annotated by associated COIL-20 object hypothesis

arrow)! To the knowledge of the authors, the entropy critical classifier is outperforming any comparable method in the literature, since comparable local recognition results have been achieved without noise only.

Outdoor Recognition Performance (TSG-20). The TSG-20 database[2] comprises images from 20 geo-referenced objects, i.e., facades of buildings from the city of Graz, Austria. For each object, we selected the frontal view for training to determine informative, and 2 images taken by a viewpoint change of $\approx \pm 30°$ for test purposes (giving 40 test images in total). Imagettes were chosen of size 21×21, and projected into 20-dimensional eigenspace, using entropy threshold $\Theta = 1.0$ (Fig. 3) for the attentive mapping and neighborhood parameter $\epsilon = 0.1$ in the resulting experiments. We compared these results with previous experiments with the TSG-I database [7]. The recognition rate on the TSG-1 database was 100%, with both $\Theta = 1.0$ and $\Theta = 3.0$, while for the TSG-20 database the recognition rate was 95% for $\Theta = 1.0$ and 97.5% for $\Theta = 2.0$. Figure 8 shows a sample recognition experiment on the TSG-20 database.

Attentive Object Detection. Preliminary experiments on the attentive object detection system have been performed to evaluate it in the presence of background clutter and noise (Fig. 7). Objects from the COIL-20 database were copied into an image of an everydays environment (a) for detection. Local appearance pattern were first interpreted in a most rapid image analysis using the rapid decision tree based estimator of the entropy function (b). Applying a threshold on the estimated entropy to obtain regions of interest for further analysis ($\Theta \leq 1.0$ was used on Fig. 7(b)). Finally, patterns inside the discriminative regions were mapped into eigenspace for a 1-NN analysis with respect to the sparse representation ($\Theta \leq 1.0$). This results in object hypotheses for nearest neighbor distances in eigenspace $\leq \epsilon = 0.1$, and no labelling otherwise. The result of the interpretation (smoothed with a Gaussian filter of $\sigma = 4.0$) is depicted in Fig. 7(c): most of the estimated object locations correspond with pixel coordinates of the associated object pattern in (a). E.g., attentive search for COIL object number 4 would result in the exact position of the queried object. As some artefacts may not correspond to object

---

[2] The TSG-20 (Tourist Sights Graz) database can be downloaded at the URL http://dib.joanneum.at/cape/TSG-20.

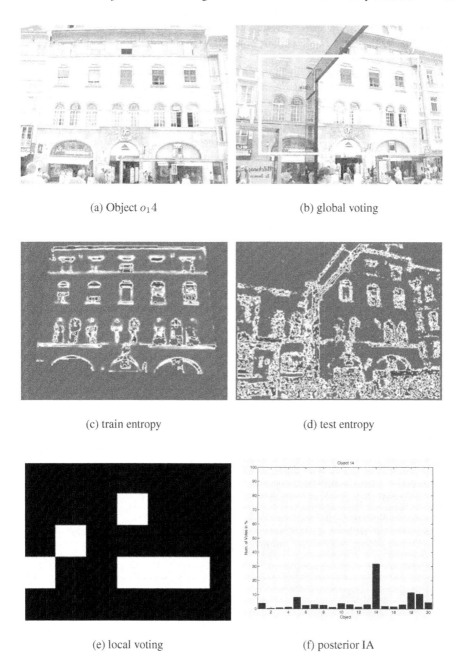

(a) Object o_14

(b) global voting

(c) train entropy

(d) test entropy

(e) local voting

(f) posterior IA

Fig. 8. Sample building recognition for object o_{14} of TSG-20 database: (a) train image, (b) test image with result overlaid, (c) entropy image of (a), entropy image of (b), (e) local voting results, and (f) posterior distribution for informative appearances (IA)

patterns, we expected to further disambiguate these locations from a more thorough analysis in a hierarchical framework of object recognition, dedicated to future work.

Discussion. The experiments illustrate that discriminative interest operators may provide a good basis for multi-stage task based attention for object recognition. Further improvements should be achieved in a hierarchical framework, integrating local with more global information, or concerning spatial correlation between object fragments [21]. Further improvements might arise from excluding patterns at object boundaries, and from reasoning across scales.

6   Conclusions

This work represents a thorough statistical analysis of local discriminative information for attentive object recognition, applied to objects from the COIL20 database within a cluttered background. It demonstrates that the local information content of an image with respect to object recognition provides a favourable measure to determine both sparse object models and interest operators for attentive detection. Focusing image analysis exclusively on discriminative regions will not only result in accelerated processing but even in superior recognition performance (Sec. 5). The methods potential for applications is in object detection tasks, such as in rapid and robust video analysis. The presented methodology represents a starting point for an extended view on image interpretation that particularly takes statistical characteristics into concern for both bottom-up and top-down information selection, dependent on a given task and its corresponding objective, such as discrimination of objects.

This work proposes a robust method, i.e., – the informative features approach – to recognize buildings in local environments, and demonstrates efficient performance on the TSG-20 database, including geo-referenced images from 20 objects in the city of Graz. This method robustly provides high recognition rates and discriminative beliefs on occluded imagery, with some degrees of viewpoint, scale, and illumination changes.

Future work will focus on finding appropriate methods to further thin out the sparse object model, and taking local topology in subspace into concern. Furthermore, ongoing work considers to develop a grouping mechanism that would locally segment promising regions of interest for detection with the goal of cascaded object detection.

Acknowledgments

This work is funded by the European Commission's project DETECT under grant number IST-2001-32157, and by the Austrian Joint Research Project on Cognitive Vision under sub-projects S9103-N04 and S9104-N04.

References

1. S. Agarwal and D. Roth. Learning a sparse representation for object detection. In *Proc. European Conference on Computer Vision*, volume 4, pages 113–130, 2002.

2. J. Braun, C. Koch, D.K. Lee, and L. Itti. Perceptual consequences of multilevel selection. In J. Braun, C. Koch, and J.L. Davies, editors, *Visual Attention and Cortical Circuits*, pages 215–241. The MIT Press, Cambridge, MA, 2001.
3. V. C. de Verdiére and J. L. Crowley. Visual recognition using local appearance. In *Proc. European Conference on Computer Vision*, 1998.
4. R. Fergus, P. Perona, and A. Zisserman. Object class recognition by unsupervised scale-invariant learning. In *Proc. IEEE Conference on Computer Vision and Pattern Recognition*, pages 264–271, 2003.
5. J. H. Friedman, J. L. Bentley, and R. A. Finkel. An algorithm for finding best matches in logarithmic expected time. *ACM Transactions on Mathematical Software*, 3(3):209–226, 1977.
6. G. Fritz, L. Paletta, and H. Bischof. Object recognition using local information content. In *Proc. International Conference on Pattern Recognition, ICPR 2004*, volume II, pages 15–18. Cambridge, UK, 2004.
7. G. Fritz, C. Seifert, L. Paletta, P. Luley, and A. Almer. Mobile vision for tourist information systems in urban environments. In *Proc. International Conference on Mobile Learning, MLEARN 2004*. Bracciano, Italy, 2004.
8. D. Hall, B. Leibe, and B. Schiele. Saliency of interest points under scale changes. In *Proc. British Machine Vision Conference*, 2002.
9. Martin Jägersand. Saliency maps and attention selection in scale and spatial coordinates: An information theoretic approach. In *Proc. International Conference on Computer Vision*, 1995.
10. T. Kadir and M. Brady. Scale, saliency and image description. *International Journal of Computer Vision*, 45(2):83–105, 2001.
11. D. Lowe. Object recognition from local scale-invariant features. In *Proc. International Conference on Computer Vision*, pages 1150–1157, 1999.
12. K. Mikolajczyk and C. Schmid. An affine invariant interest point detector. In *Proc. European Conference on Computer Vision*, pages 128–142, 2002.
13. H. Murase and S. K. Nayar. Visual learning and recognition of 3-D objects from appearance. *International Journal of Computer Vision*, 14(1):5–24, 1995.
14. S. Obdrzalek and J. Matas. Object recognition using local affine frames on distinguished regions. In *Proc. British Machine Vision Conference*, pages 113–122, 2002.
15. L. Paletta and C. Greindl. Context based object detection from video. In *Proc. International Conference on Computer Vision Systems*, pages 502–512, 2003.
16. E. Parzen. On estimation of a probability density function and mode. *Annals of Mathematical Statistics*, 33:1065–1076, 1962.
17. J.R. Quinlan. *C4.5 Programs for Machine Learning*. Morgan Kaufmann, San Mateo, CA, 1993.
18. N. Saito, R.R.Coifman, F.B. Geshwind, and F. Warner. Discriminant feature extraction using empirical probability density estimation and a local basis library. volume 35, pages 2841–2852, 2002.
19. D.L. Swets and J. Weng. Using discriminant eigenfeatures for image retrieval. *IEEE Transactions on Pattern Analysis and Machine Intelligence*, 18(8):831–837, 1996.
20. P. van de Laar, T. Heskes, and S. Gielen. Task-dependent learning of attention. *Neural Networks*, 10(6):981–992, 1997.
21. M. Vidal-Naquet and S. Ullman. Object recognition with informative features and linear classification. In *Proc. International Conference on Computer Vision*, 2003.
22. M. Weber, M. Welling, and P. Perona. Unsupervised learning of models for recognition. In *Proc. European Conference on Computer Vision*, pages 18–32, 2000.

A Model of Object-Based Attention That Guides Active Visual Search to Behaviourally Relevant Locations

Linda Lanyon and Susan Denham

Centre for Theoretical and Computational Neuroscience,
University of Plymouth, U.K
linda.lanyon@plymouth.ac.uk

Abstract. During active visual search for a colour-orientation conjunction target, scan paths tend to be guided to target coloured locations (Motter & Belky, 1998). An active vision model, using biased competition, is able to replicate this behaviour. At the cellular level, the model replicates spatial and object-based attentional effects over time courses observed in single cell recordings in monkeys (Chelazzi et al., 1993, 2001). The object-based effect allows competition between features/objects to be biased by knowledge of the target object. This results in the suppression of non-target features (Chelazzi et al., 1993, 2001; Motter, 1994) in ventral "what" stream areas, which provide a bias to the spatial competition in posterior parietal cortex (LIP). This enables LIP to represent behaviourally relevant locations (Colby et al., 1996) and attract the scan path. Such a biased competition model is extendable to include further "bottom-up" and "top-down" factors and has potential application in computer vision.

1 Introduction

The biased competition hypothesis [1][2][3][4] is currently very influential in the study of visual attention at many levels ranging from neurophysiological single cell studies to brain imaging, evoked potentials and psychophysics. The hypothesis suggests that responses of neurons, such as those that encode object features, are determined by competitive interactions. This competition is subject to a number of biases, such as "bottom-up" stimulus information and "top-down" cognitive requirements, for example a working memory template of the target object during visual search. Building on early small-scale models using this hypothesis, e.g.[5][6], systems level models, e.g. [7][8][9], have been able to replicate a range of experimental data. However, there has been no systems level modelling of the biased competition hypothesis for active visual search, where retinal inputs change as the focus of attention is shifted. The model presented here uses biased competition and adopts an active vision approach such that, at any particular fixation, its cortical modules receive retinal inputs relating to a portion of the entire scene. This allows the model to process a smaller area of the image at any fixation and provides a realistic model of normal everyday search.

Traditionally, visual attention was thought to operate as a simple spatial "spotlight" [10][11][12] that acted to enhance responses at a particular location(s) and it is clear that attention can produce spatially specific modulation of neuronal responses, e.g.[13][14],

L. Paletta et al. (Eds.): WAPCV 2004, LNCS 3368, pp. 42–56, 2005.

and behaviour [15]. However, many studies in psychophysics [16][17], functional magnetic resonance imaging [18], event-related potential recordings [19][20] and single cell recordings [21][22][23] provide convincing evidence for attention operating in a more complex object-based manner. However, these findings are confounded by issues relating to whether the selection is object-based, feature-based or surface-based. Here we use the term object-based in the sense that it refers an attentional modulation of cellular responses on the basis of the features of a target object, as used by [7][21][23]. The term feature-based is also appropriate. Many different mechanisms may be responsible for the range of object-based effects seen experimentally and not all are addressed here. For example, there is no attempt to provide Gestalt-like perceptual grouping, but an extension to the model's feature processing modules to allow local lateral connectivity across similar features could produce such an effect.

Event-related potential [24] and single cell recordings suggest that spatial attention may be able to modulate responses earlier than object-based attention. For example, in monkey lateral intraparietal area (LIP), an early spatial enhancement of responses has been recorded from single cells in anticipation of the stimulus [25] and has been seen in imaging of the possible human homologue of LIP [26][27][28]. In ventral pathway area V4, which encodes form and colour features [29], spatial modulation of baseline responses has been recorded from single cells in advance of the sensory response and the earliest stimulus-invoked response at 60ms post-stimulus was also spatially modulated [14]. However, object-based effects in V4 appear to take longer to be resolved and have been recorded from approximately 150ms post-stimulus [23][30][31]. Such modulation, leading to the response of the cell being determined by which object within its receptive field is attended, is known as the 'target effect' [23]. In inferior temporal cortex (IT), object-based modulation of responses during a delay period following a cue stimulus has been recorded prior to the onset of the response relating to the search array [32]. However, the initial sensory response invoked by the search array is generally not significantly modulated by the target object compared to the strong target effect developing from 150-200ms post-onset [21][32]. The model presented here replicates these spatial and object-based effects, at the neuronal level, with temporal precision. Such effects at the neuronal level allow the model, at the systems level, to produce the search behaviour similar to that seen in humans [33][34] and monkeys [35].

A "cross-stream" interaction between the ventral "what" pathway, responsible for the encoding of features and objects and leading from primary visual cortex (V1) through V2 and V4 to temporal cortex, and the dorsal "where" pathway, leading from V1 to parietal cortex [36][37], allows the search scan path to be drawn to behaviourally relevant locations rather than less relevant stimulus locations or blank areas. Object-based attention in the model's ventral pathway biases the dorsal module (LIP) to represent the location of target features most strongly [25][38]. In addition, this connection allows certain features to have priority in attracting attention. In particular, the model is able to reproduce the scan path behaviour of monkeys searching for a colour-orientation feature conjunction target [35] where fixations tended to land within 1° of stimuli (only 20% fell in blank areas of the display) and these stimuli tended to be target coloured (75% of fixations landed near target coloured stimuli and only 5% near non-target coloured stimuli). The use of object-based attention within the model to guide the scan path is the focus of this paper.

2   The Model

Fig. 1 shows a schematic overview of the model, which is formally described in [39]. It consists of a retina and V1, which detect form and colour features in a retinotopic manner, and three dynamic modules representing areas V4, IT and LIP, which are modelled using a mean field approach [40], in a similar manner to the model in [7], where the level of representation is a population of cells with similar properties, known as a cell assembly. The size of the retina can be varied for any particular simulation because the size of the cortical modules is scaled according to retina size. Thus, a larger retina may be used where biologically valid predictions are to be made, or a smaller retina where processing speed needs to be optimised (the only restriction being that a very small retina tends to lead to a higher proportion of fixations landing in blank areas of very sparse scenes due to lack of stimuli within the limited retina). For reasons of computational simplicity and speed, V1 is not included in the dynamic portion of the system and there is

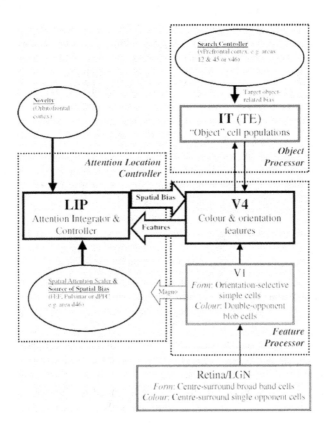

Fig. 1. Overview of the model

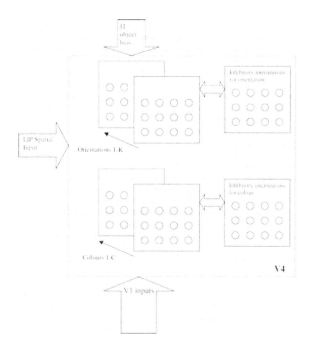

Fig. 2. Competition in V4

no attempt in this version of the model to reproduce attentional effects therein, although such effects have been observed [22][41]. Intermediate stages of visual processing, such as area V4, are the focus of this model. The V1 module contains a neuron at every pixel location and features detected in V1 are fed forward to V4 in a convergent manner over the V4 receptive field area. Thus, V4 is retinotopic with biologically plausible [42] larger receptive fields than found in V1 (the size of filters used in V1 determines the ratio of pixels to degrees of visual angle). Stimuli are coloured oriented bars, chosen to match the feature conjunctions used by [35], in order to reproduce the effects seen in that psychophysical experiment. V4 neurons are known to be functionally segregated [43] and the V4 module encodes colour and orientation in separate feature layers, a common modelling simplification. Different features of the same type at the same location compete via inhibitory interneuron assemblies (i.e. different colours within the same V4 receptive field area compete and different orientations within the same V4 receptive field area compete), as shown in fig. 2. This provides the necessary competition between features within the same receptive field that has been found in single cell recordings [44][14][21][23]. V4 featural information is fed forward to activate non-retinotopic invariant object populations in IT. Thus, the model's ventral pathway (V1, V4 and IT) operates as a feature/object processing hierarchy with receptive field sizes increasing towards IT, where receptive fields span the entire retina. It is known that anterior areas of IT, such as area TE, are not retinotopic but have large receptive fields and encode objects in an invariant manner [42].

Dorsal pathway module LIP contains a cell assembly for every retinotopic location in V4. Each LIP assembly is reciprocally connected to the assembly in each feature layer in V4 at that location. LIP provides a spatio-featural map that is used to control the spatial focus of attention and, hence, fixation. Although responses in monkey LIP have been found to be independent of motor response [45], LIP is thought to be involved in selecting possible targets for saccades and has connections to superior colliculus and the frontal eye field (FEF), which are involved in saccade generation. Furthermore, direct electrical stimulation of LIP has elicited saccades [46]. Competition in the model LIP is spatial and the centre of the receptive field of the most active assembly at the end of the fixation is chosen as the next fixation point. During fixation a spatial bias creates an early spatial attention window (AW), which is scaled according to local stimulus density (on the basis of low resolution orientation information assumed to be conveyed rapidly to parietal cortex through the magnocellular pathway), as described by [47]. This reflects the scaling of the area around fixation within which targets can be detected [48]. This spatial bias to LIP creates an anticipatory spatial attention effect, as has been found in single cell recordings in this area [25]. LIP provides a spatial bias to V4 and this creates an early spatial attention effect therein, modulating baseline firing before the onset of the stimulus evoked response and, subsequently, modulating this response from its onset at $\sim$60ms post-stimulus, as found by [14]. The source of the spatial bias could be FEF because stimulation of this area leads to a spatially specific modulation of neuronal response in V4 [49]. The spatial bias from LIP to V4 also provides some binding across feature types in V4.

Within V4, the early spatial focus of attention becomes object-based over time. This is due to the fact that V4 is subject to two concurrent biases: A spatial bias from LIP and an object-related bias from IT. Thus, object-based attention evolves in the model's ventral pathway due to its object-related feedback biases. The development of object-based attention in V4 is dependent on the resolution of competition between objects in IT. In addition to the feed forward inputs from V4, competition between objects in IT is biased by a feedback current from prefrontal cortex, which is assumed to hold a working memory template of the target object. IT feeds back an inhibitory bias to V4 that suppresses features not related to the object. Over time, the target object is able to win the competition in IT, and then target features become enhanced whilst non-target features are suppressed across V4, as has been found in single cell recordings in monkeys [30][31] over the same time course. Once a significant object-based effect is established in IT, a saccade is initiated 70ms later, to reflect motor preparation, and replicate the timing of saccade onset following the target object effect seen in single cell recordings in IT [21] and V4 [23].

Following this object-based "highlighting" of target features in parallel across V4 [30][31], the featural inputs from V4 to LIP provide a bias that results in locations containing target-related features becoming most active in LIP. Thus, LIP is able to represent the locations of salient, behaviourally relevant stimuli [25][38]. Prior to the onset of object-based attention in the ventral stream (which occurs $\sim$150ms post-stimulus, as found by [21][23][32], the initial sensory response in LIP tends to represent locations containing stimuli equally, but more strongly than blank areas. The connection from V4 to LIP is set such that colour features provide a slightly stronger bias than orientation

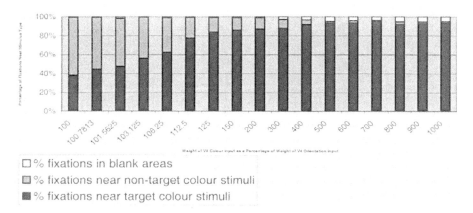

% fixations in blank areas
% fixations near non-target colour stimuli
% fixations near target colour stimuli

Fig. 3. The effect of increasing the relative weight of V4 colour feature input to LIP. When V4 colour features are marginally more strongly connected to LIP than V4 orientation features, the scan path is attracted to target coloured stimuli in preference to stimuli of the target orientation

features. Thus, LIP tends to represent locations containing target coloured stimuli more strongly than locations containing the target orientation but non-target colour. Therefore, colour is given a priority in guiding the scan path, as found by [35]. The difference in strength of connection of the V4 features to LIP need only be marginal in order to achieve this effect. Fig. 3 shows the effect of adjusting the relative connection weights.

The use of this active vision paradigm with a moving retina means that inhibition of return (IOR) in the scan path cannot be implemented by inhibition of previously active cortical locations in retinotopic coordinates. Parietal damage is linked to an inability to retain a spatial working memory of searched locations across saccades so that locations are repeatedly re-fixated [50]. Here, IOR in the scan path is provided by a novelty-related bias to LIP. The source of such a bias could be a frontal area, such as orbitofrontal cortex, which is linked with event/reward associations and, when damaged, affects IOR [51]. A world/head-based map reflects the novelty of all locations in the scene and, hence, their potential reward. Initially, the novelty of all locations in the scene is high. Then, when attention is withdrawn from a location, locations in the immediate vicinity of the fixation point have their novelty reduced to a low value that climbs in a Gaussian fashion with distance from fixation such that peripheral locations within the AW have neutral novelty. Novelty recovers over time such that IOR is present at multiple locations [52][53][54], the magnitude of the effect decreases approximately linearly from its largest value at the most recently searched location and several previous locations are affected [55][56].

3   Results

3.1   Neuronal Level Results

At the neuronal level, attentional effects within V4 evolve from being purely spatial to being object-based, over time courses found in single cell recordings from monkeys [14][23][30][31]. The enhancement of target features across V4 is particularly important to the guidance of the scan path and fig. 4 shows the development of object-based

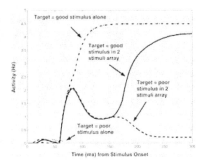

Fig. 4. Comparison of V4 single assembly response: Shows the replication of [23] figure 5 from array onset until the time of the saccade. Here, saccades have been suppressed but, for the two stimuli case, would have occurred at ~235-240ms post-stimulus, the same onset time as found by [23]. A good stimulus for the cell (i.e. one that causes a high response) presented alone in the receptive field produces a high response: Top line. When a poor stimulus is presented alone the cell's response is suppressed: Bottom line. When both stimuli are simultaneously presented within the receptive field, the high response normally invoked by the good stimulus is reduced due to the competing presence of the poor stimulus. However, once object-based effects begin from ~150ms, the response is determined by which of the two stimuli is the target, with activity tending towards that of the single stimulus case for the target object. The 2 stimuli case when the target is a poor stimulus for the cell shows the suppression of the non-target responses (i.e. the cell's response to the good stimulus) that enables V4 to represent target stimuli most strongly and bias LIP to guide the scan path to these locations

attention from ~150ms post-stimulus in V4, which mirrors that recorded by [23]. This effect in V4 is due to object-related feedback in IT and depends on the development of the target object effect in IT, which is shown in fig. 5. Fig. 5 replicates the time course of the development of a significant target object effect in the sensory response, as found by [21] and [32]. Figs. 4 and 5 replicate data from the onset of the sensory response (the start of the fixation in the model) and do not address the maintenance of cue-related activity during a delay period, which is found in IT [21][32]. In some IT cells there is a slight modulation of early sensory responses on the basis of the target object [32]. However, this is insignificant compared to the target object effect later in the response and has been suggested [32] to be due to a continuation of elevated firing during the delay interval. Although, sustained activity in IT during a delay period is not specifically modelled here, it is possible to reproduce a weak early target object modulation, in addition to the significant target effect beginning ~150ms post-stimulus. The figure on the right of fig. 5 shows this effect and is produced by using a slightly more complex form of prefrontal feedback to IT (combining a small constant excitatory feedback, representing sustained delay period activity, with a climbing inhibitory feedback). Further discussion about the replication of both spatial and object-based attentional effects, and accurate onset timing, from monkey single cell data in IT [21][32] and V4 [14][23] has been reported in more detail elsewhere and is beyond the scope of this paper. However, replication of data from older and more highly trained monkeys [23] compared to [21] suggested that IT feedback to V4 might be tuned by learning.

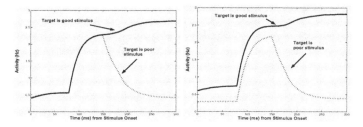

Fig. 5. Target object effect in IT: The plot on left shows the replication of [21] figure 3a from array onset until the time of the saccade. Two stimuli (a good stimulus for the cell and a poor one) are simultaneously presented within the cell's receptive field. The initial sensory response is not affected by target object selection but from ∼150ms post-stimulus the response is determined by which of the two objects is the search target. If the target is the poor stimulus, responses are significantly suppressed. The plot on the right shows the same situation but includes a slight target effect early in the sensory response, as found in some cells. This replicates [32] figure 7a from array onset until the time of the saccade. A more complex prefrontal feedback to IT, including a sustained excitatory component, is used here. As in the left-hand plot, the significant target object effect begins ∼150ms post-stimulus

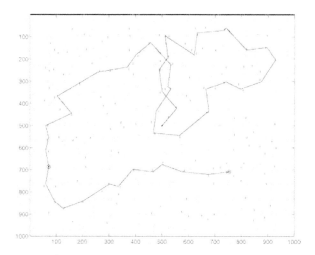

Fig. 6. Scan path through a dense scene: The target is a red stimulus. Fixations are shown as i) magenta dots - within $1°$ of a target coloured stimulus; 96% of fixations; ii) blue circles - within $1°$ of a non-target colour stimulus; 4% of fixations. Average saccade amplitude = $7.4°$. (N.B. When figures containing red and green bars are viewed in greyscale print, the red appears as a darker grey than the green)

3.2   Scan Path Results

As found in a similar psychophysical task [35], scan paths are attracted to target coloured locations. This is shown in fig. 6 where the target is a red horizontal bar and fig. 7, where the target is a green horizontal bar. Saccades tend to be shorter in dense scenes compared to sparse scenes. This is because the AW is scaled according to stimulus density and it

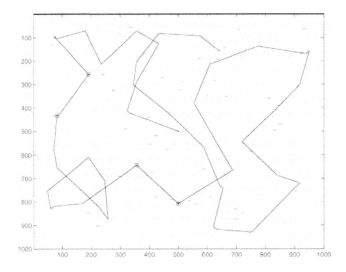

Fig. 7. Scan path through a sparse scene: The target is a green stimulus. Fixations are shown as i) magenta dots- within 1 ° of a target coloured stimulus; 92% of fixations; ii) blue circles - within 1 ° of a non-target colour stimulus; 8% of fixations. Average saccade amplitude = 12.1 °

contributes a positive bias to the competition in LIP, meaning that next fixation points tend to be chosen from within the AW. The AW scaling is based on local stimulus density. Therefore, the AW expands and contracts as it moves around a scene of mixed density, resulting in smaller saccades in the dense areas and larger saccades in sparse areas. Updates to the novelty map are also based on the size of the AW and this means that a smaller region is inhibited for return in areas of dense stimuli. Thus, the scan path is able to investigate the dense, and potentially more interesting, areas of a scene with a series of shorter amplitude saccades, as shown in fig. 8. This is potentially useful for natural scene processing, where uniform areas such as sky could be examined quickly with large amplitude saccades, allowing the scan path to concentrate in more detailed and interesting aspects of the scene.

Increasing the weight of the novelty bias to LIP slightly reduces the likelihood of fixating target coloured stimuli and, in sparse scenes, increases the number of fixations landing in blank areas of the display. Fig. 9 shows this effect. Thus, a search where novelty is the key factor, for example a hasty search of the entire scene due to time constraints, may result in more "wasted" fixations in blank areas.

Weaker object-related feedback within the ventral pathway (prefrontal to IT; IT to V4) reduces the object-based effect within V4 and this results in more non-target coloured stimuli being able to capture attention. Saccade onset is also delayed. The effect of the weight of IT feedback to V4 on fixation position is shown in fig. 10. Replication of single cell data suggested that IT feedback to V4 could be tuned by learning and experience. Therefore, search during a familiar task or with familiar objects may be faster and the scan path would be expected to fixate more target coloured locations than when the task or objects were unfamiliar.

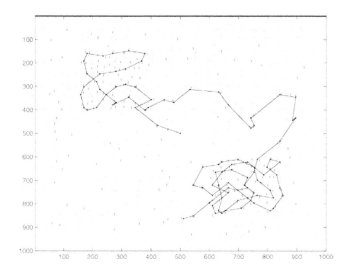

Fig. 8. Scan path through a mixed density scene: the scan path examines the dense areas more thoroughly with a series of shorter saccades (average amplitude 4.5 ° in dense areas, compared to 7.5 ° in slightly more sparse areas)

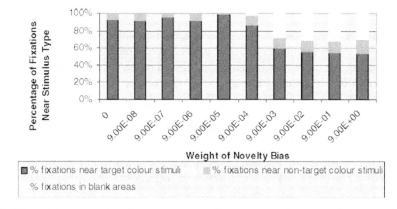

Fig. 9. Effect of the novelty bias in a sparse scene: Shows average fixation positions over 10 scan paths, each consisting of 50 fixations, over the image shown in fig. 7. Fixations are considered to be near a stimulus when they are within 1 ° of a stimulus centre. As the weight of the novelty bias increases, the number of fixations near target coloured stimuli decreases and fixations are more likely to occur in blank areas of the display

4   Discussion

This model is able to reproduce data from monkeys at both the neuronal [14][21] [23][25][30][31][32] and behavioural [35][48] levels. At the neuronal level, this is the first model to show attentional effects within V4 evolving from being purely spatial to being object-based, over time courses found in single cell recordings from monkeys

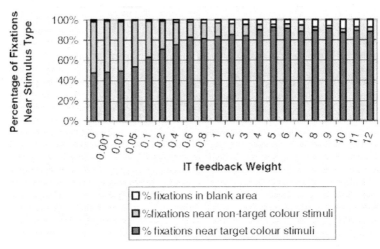

Fig. 10. Effect of weight of IT feedback to V4: Shows average fixation positions over 10 scan paths, each consisting of 50 fixations, over the image shown in fig. 6. Fixations are considered to be near a stimulus when they are within 1 ° of a stimulus centre. As the weight of feedback is increased there is a tendency for more target coloured stimuli and less non-target coloured stimuli to be fixated because object-based effects in the ventral stream are stronger

[14][23]. Object-based effects in IT also accurately reproduce those recorded in monkey IT during visual search [21][32]. Single cell recordings from monkeys [21][23] suggest that saccade onset may be temporally linked to the development of significant object-based effects in ventral pathway cortical areas. Here, saccade onset is linked to the development of a significant effect in IT. This leads to saccade onset times that are able to replicate those found by [21][23].

Of potential future use in practical applications such as video surveillance and robot vision is the ability to use these object-based effects to guide an active vision scan path. As a result of the "top-down" cognitive biasing of competition within the model (by prefrontal feedback to the model's ventral pathway), saccades are guided to behaviourally relevant locations that contain a potential target stimulus. Thus, object-based attention (in the ventral pathway) is able to bias spatial competition (within the dorsal pathway), enabling the LIP module to be a spatio-featural integrator that can control the scan path. The model is extendable to include other factors in the competition in LIP. For example, further stimulus-related factors, such as luminance or motion, could contribute to increase the salience of a location. Thus, the model is extendable to include other endogenous and exogenous factors in the competition for the capture of attention; an issue currently more widely explored in psychophysics, e.g. [57], than modelling, but of much practical relevance in future computer vision.

Selectivity in the guidance of eye movements during visual search, giving a preference for locations containing target features, has been found in a number of studies (e.g. [33][34][35][58]; but see [59]). Under several conditions with simple stimuli, colour (or luminance) provides stronger guidance than orientation in monkeys [35] and humans [33][34] (but see [58] for equal preference for shape and colour). At the behavioural level, the model is able to prioritise the features that guide the scan path, such that this tendency for colour to dominate search can be reproduced. The priority of colour in guiding the scan path is achieved by a marginal difference in the relative strength of connection from V4 feature types to LIP. The relative strength of these connections may be malleable on the basis of cognitive requirement or stimulus-related factors, such as distractor ratios [60]. However, under conditions of equal proportions of distractor types, colour was found to guide the scan path even when the task was heavily biased towards orientation discrimination [35]. This suggests that such connections may be fixed during development and may not be easily adapted to task requirements. The model highlights the possible role for feedforward connections from the ventral stream to parietal cortex in determining search selectivity, and the need for further investigation of such connections in lesion studies. The model makes further predictions beyond the scope of this paper relating to the effect of damage to the LIP/V4 connections on binding of information across feature dimensions.

The requirement for a novelty map in world/head-based co-ordinates as a memory trace of locations visited suggests that visual search cannot be totally amnesic [61] and supports the idea of memory across saccades [56][62][63]. The involvement of frontal and parietal areas in this form of IOR may explain why orbitofrontal and parietal patients display increased revisiting of locations in their scan paths [50][51].

The ability of this model to reproduce active visual search behaviour found in humans and monkeys whilst accurately replicating neurophysiological data at the single cell and brain region levels, suggests that biological models can lend much to active vision research for practical computer vision applications.

References

1. Duncan, J., Humphreys, G.W.: Visual search and stimulus similarity. Psych. Rev. Vol. 96(3) (1989) 433–458
2. Desimone, R., Duncan, J.: Neural mechanisms of selective visual attention. Ann. Rev. Neuroscience. Vol. 18 (1995) 193–222
3. Duncan, J., Humphreys, GW., Ward, R.: Competitive brain activity in visual attention. Current Opinion in Neurobiology. Vol. 7 (1997) 255–261
4. Desimone, R.: Visual attention mediated by biased competition in extrastriate visual cortex. Phil. Trans. Royal Soc. London B. Vol. 353 (1989) 1245–1255
5. Reynolds, J.H., Chelazzi, L., Desimone, R.: Competitive mechanisms subserve attention in macaque areas V2 and V4. J. Neuroscience. Vol. 19(5) (1999) 1736–1753
6. Usher, M., Niebur, E.: Modeling the temporal dynamics of IT neurons in visual search: A mechanism for top-down selective attention. J. Cognitive Neuroscience. Vol. 8(4) (1996) 311–327
7. Deco, G., Lee, T.S.: A unified model of spatial and object attention based on inter-cortical biased competition. Neurocomputing. Vol. 44-46 (2002) 775–781
8. De Kamps, M., Van der Velde, F.: Using a recurrent network to bind form, color and position into a unified percept. Neurocomputing. Vol. 38-40 (2001) 523–528
9. Hamker, F.H.: A dynamic model of how feature cues guide spatial attention. Vision Research. Vol. 44 (2004) 501–521
10. Helmholtz, H: Handbuch der physiologishen Optik, Voss, Leipzig (1867)
11. Treisman, A.: Perceptual grouping and attention in visual search for features and for objects. J. Exp. Psych.: Human Perc. & Perf. Vol. 8 (1982) 194–214

12. Crick, F.: Function of the thalamic reticular complex: The searchlight hypothesis. Proc. Nat. Ac. Sci., USA. Vol. 81 (1984) 4586-4590

13. Connor, C.E., Callant, J.L., Preddie, D.C., Van Essen, D.C.: Responses in area V4 depend on the spatial relationship between stimulus and attention. J. Neurophysiology. Vol. 75 (1996) 1306–1308

14. Luck, S.J., Chelazzi, L., Hillyard, S.A., Desimone, R.: Neural mechanisms of spatial attention in areas V1, V2 and V4 of macaque visual cortex. J. Neurophysiology. Vol. 77 (1997) 24–42

15. Bricolo, E., Gianesini, T., Fanini, A., Bundesen, C., Chelazzi, L.: Serial attention mechanisms in visual search: a direct behavioural demonstration. J. Cognitive Neuroscience. Vol. 14(7) (2002) 980–993

16. Duncan, J.: Selective attention and the organisation of visual information. J. Exp. Psych.: General. Vol. 113 (1984) 501–517

17. Blaser, E., Pylyshyn, Z.W., Holcombe, A.O.: Tracking an object through feature space. Nature. Vol. 408(6809) (2000) 196–199

18. O'Craven, K.M., Downing, P.E., Kanwisher, N.: fMRI evidence for objects as the units of attentional selection. Nature. Vol. 401 (1999) 584–587

19. Valdes-Sosa, M., Bobes, M.A., Rodriguez, V., Pinilla, T.: Switching attention without shifting the spotlight: Object-based attentional modulation of brain potentials. J. Cognitive Neuroscience. Vol. 10(1) (1998) 137–151

20. Valdes-Sosa, M., Cobo, A., Pinilla, T.: Attention to object files defined by transparent motion. J. Exp. Psych.: Human Perc. & Perf. Vol. 26(2) (2000) 488–505

21. Chelazzi, L., Miller, E.K., Duncan, J., Desimone, R.: A neural basis for visual search in inferior temporal cortex. Nature. Vol. 363 (1993) 345–347

22. Roelfsema, P.R., Lamme, V.A.F., Spekreijse, H.: Object-based attention in the primary visual cortex of the macaque monkey. Nature. Vol. 395 (1998) 376–381

23. Chelazzi, L., Miller, E.K., Duncan, J., Desimone, R.: Responses of neurons in macaque area V4 during memory-guided visual search. Cerebral Cortex. Vol. 11 (2001) 761–772

24. Hillyard, S.A., Anllo-Vento, L.: Event-related brain potentials in the study of visual selective attention, Proc. Nat. Acad, Sci. USA. Vol. 95 (1998) 781–787

25. Colby, C.L., Duhamel, J.R., Goldberg, M.E.: Visual, presaccadic, and cognitive activation of single neurons in monkey lateral intraparietal area. J. Neurophysiology. Vol. 76(5) (1996) 2841–2852

26. Kastner, S., Pinsk, M., De Weerd, P., Desimone, R., Ungerleider, L.: Increased activity in human visual cortex during directed attention in the absence of visual stimulation. Neuron. Vol. 22 (1999) 751–761

27. Corbetta, M., Kincade, J.M., Ollinger, J.M., McAvoy, M.P., Shulman, G.L.: Voluntary orienting is dissociated from target detection in human posterior parietal cortex. Nature Neuroscience Vol. 3 (2000) 292–297

28. Hopfinger, J.B., Buonocore, M.H., Mangun, G.R.: The neural mechanisms of top-down attentional control. Nature Neuroscience. Vol. 3(3) (2000) 284–291

29. Zeki, S.: A Vision of the Brain, Blackwell Scientific Publications, Oxford, UK. (1993)

30. Motter, B.C.: Neural correlates of attentive selection for color or luminance in extrastriate area V4. J. Neuroscience. Vol. 14(4) (1994) 2178–2189

31. Motter, B.C.: Neural correlates of feature selective memory and pop-out in extrastriate area V4. J. Neuroscience. Vol 14(4) (1994) 2190–2199

32. Chelazzi, L., Duncan, J., Miller, E.K.: Responses of neurons in inferior temporal cortex during memory-guided visual search. J. Neurophysiology. Vol 80 (1998) 2918–2940

33. Scialfa, C.T., Joffe, K.M.: Response times and eye movements in feature and conjunction search as a function of target eccentricity. Perception & Psychophysics. Vol. 60(6) (1998) 1067–1082

34. Williams, D.E., Reingold, E.M.: Preattentive guidance of eye movements during triple conjunction search tasks: The effects of feature discriminability and saccadic amplitude. Psychonomic Bulletin & Review. Vol. 8(3) (2001) 476–488
35. Motter, B.C., Belky, E.J.: The guidance of eye movements during active visual search. Vision Research. Vol. 38(12) (1998) 1805–1815
36. Ungerleider, L.G., Mishkin, M.: Two cortical visual systems. In: Ingle, D.J., Goodale, M.A., Mansfield, R.W.J. (eds.): Analysis of Visual Behaviour. MIT Press, MA (1982) 549–586
37. Milner, A.D., Goodale, M.A.: The Visual Brain In Action. Oxford University Press (1995)
38. Gottlieb, J.P., Kusunoki, M., Goldberg, M.E.: The representation of visual salience in monkey parietal cortex. Nature. Vol. 391 (1998) 481–484
39. Lanyon, L.J., Denham, S.L.: A Model of Active Visual Search with Object-Based Attention Guiding Scan Paths. Neural Networks, Special Issue: Vision and Brain. Vol. 17 (2004) 873–897
40. Gerstner, W.: Population dynamics of spiking neurons: Fast transients, asynchronous states, and locking. Neural Computing. Vol. 12 (2000) 43–89
41. Brefczynski, J., DeYoe, E.A.: A physiological correlate of the 'spotlight' of visual attention. Nature Neuroscience. Vol. 2 (1999) 370-374
42. Wallis, G., Rolls, E.T.: Invariant face and object recognition in the visual system. Prog. Neurobiology. Vol. 51 (1997) 167–194
43. Ghose, G.M., Ts'O, D.Y.: Form processing modules in primate area V4. J. Neurophysiology. Vol. 77(4) (1997) 2191–2196
44. Moran, J., Desimone, R.: Selective attention gates visual processing in the extrastriate cortex. Science. Vol. 229 (1985) 782–784
45. Bushnell, M.C., Goldberg, M.E., Robinson, D.L.: Behavioural enhancement of visual responses in monkey cerebral cortex. I. Modulation in posterior parietal cortex related to selective visual attention. J. Neurophysiology. Vol. 46(4) (1981) 755–772
46. Their, P., Andersen, R.A.: Electrical microstimulation distinguishes distinct saccade-related areas in the posterior parietal cortex. J. Neurophysiology. Vol. 80 (1998) 1713–1735
47. Lanyon, L. J., Denham, S. L.: A biased competition computational model of spatial and object-based attention mediating active visual search. Neurocomputing. Vol 58–60C (2004) 655–662
48. Motter, B.C., Belky, E.J.: The zone of focal attention during active visual search", Vision Research. Vol. 38(7) (1998) 1007–1022
49. Moore, T., Armstrong, K.M.: Selective gating of visual signals by microstimulation of frontal cortex. Nature. Vol. 421 (2003) 370–373
50. Husain, M., Mannan, S., Hodgson, T., Wojciulik, E., Driver, J., Kennard, C. Impaired spatial working memory across saccades contributes to abnormal search in parietal neglect. Brain. Vol. 124 (2001) 941–952
51. Hodgson, T.L., Mort, D., Chamberlain, M.M., Hutton, S.B., O'Neill, K.S., Kennard, C.: Orbitofrontal cortex mediates inhibition of return. Neuropsychologia. Vol. 431 (2002) 1–11
52. Tipper, S.P., Weaver, B., Watson, F.L.: Inhibition of return to successively cued spatial locations: Commentary on Pratt and Abrams (1995). J. Exp. Psych.: Human Perc. & Perf. Vol. 22(5) (1996) 1289–1293
53. Danziger, S., Kingstone, A., Snyder, J.J.: Inhibition of return to successively stimulated locations in a sequential visual search paradigm. J. Exp. Psych.: Human Perc. & Perf. Vol. 24(5) (1998) 1467–1475
54. Snyder, J.J., Kingstone, A.: Inhibition of return at multiple locations in visual search: When you see it and when you don't. Q. J. Exp. Psych. A: Human Exp. Psych. Vol. 54(4) (2001) 1221–1237
55. Snyder, J.J., Kingstone, A.: Inhibition of return and visual search: How many separate loci are inhibited? Perception & Psychophysics. Vol. 62(3) (2000) 452–458

56. Irwin, D.E., Zelinsky, G.J.: Eye movements and scene perception: Memory for things observed. Perception & Psychophysics. Vol. 64(6) (2002) 882–895

57. Kim, M-S., Cave, K.R.: Top-down and bottom-up attentional control: On the nature of interference from a salient distractor. Perception & Psychophysics. Vol. 61(6) (1999) 1009–1023

58. Findlay, J.M.: Saccade target selection during visual search. Vision Research, Vol. 37(5) (1997) 617–631

59. Zelinski, G.J.: Using eye saccades to assess the selectivity of search movements. Vision Research. Vol. 36(14) (1996) 2177–2187

60. Shen, J., Reingold, E.M., Pomplum, M.: Distractor ratio influences patterns of eye movements during visual search. Perception. Vol. 29(2) (2000) 241–250

61. Horowitz, T.S., Wolfe, J.M.: Visual search has no memory. Nature. Vol. 394 (1998) 575–577

62. Mitchell, J., Zipser, D.: A model of visual-spatial memory across saccades. Vision Research. Vol. 41 (2001) 1575–1592

63. Dodd, M.D., Castel, A.D., Pratt, J.: Inhibition of return with rapid serial shifts of attention: Implications for memory and visual search. Perception & Psychophysics. Vol. 65(7) (2003) 1126–1135

Learning of Position-Invariant Object Representation Across Attention Shifts

Muhua Li and James J. Clark

Centre for Intelligent Machines, McGill University,
3480 University Street Room 410, Montreal, Quebec Canada H3A 2A7
{limh, clark}@cim.mcgill.ca

Abstract. Selective attention shift can help neural networks learn invariance. We describe a method that can produce a network with invariance to changes in visual input caused by attention shifts. Training of the network is controlled by signals associated with attention shifting. A temporal perceptual stability constraint is used to drive the output of the network towards remaining constant across temporal sequences of attention shifts. We use a four-layer neural network model to perform the position-invariant extraction of local features and temporal integration of attention-shift invariant presentations of objects. We present results on both simulated data and real images, to demonstrate that our network can acquire position invariance across a sequence of attention shifts.

1 Introduction

A computer visual system with invariant object representation should have consistent neural response, within a certain range, to the same object feature under different conditions, such as at different retinal positions or with different appearance due to viewpoint changes. Most research work done so far has focused on achieving different degrees of invariance based only on the sensory input, while ignoring the important role of visual-related motor signals. However, retrieving visual information solely on the input images can cause ill-posed problems. And the extraction of invariant features from visual images which contain overwhelming information is usually slow.

From the viewpoint of active computer vision, an effective visual system can selectively obtain useful visual information with the cooperation of motor actions. In our opinion, visual-related self-action signals are crucial in learning spatial invariance, as they provide information as to the nature of changes in the visual input. Especially, selective attention shifts could play an important role in the visual systems as to focus on a small fraction of the total input visual information ([6], [9]), to perform visual related tasks such as pattern and object recognition, as implemented in several computational models ([1], [8], [12], and [13]). Shifting of attention enables the visual system to actively and efficiently acquire useful information from the external environment for further processing. Therefore, it is conceivable to hypothesize that the learning of invariant presentation of an object might not need complete visual information about the object, instead it can be learned from those attended local features of the object across a sequence of attention shifts.

L. Paletta et al. (Eds.): WAPCV 2004, LNCS 3368, pp. 57–70, 2005.

Here we will focus on the learning of position invariance across attention shifts, with position-related retinal distortions taken into consideration. The need for developing position-invariance arises due to projective distortions and the non-uniform distribution of visual sensors on the retinal surface. These factors result in qualitatively different signals when object features are projected onto different positions of the retina. In order to produce position invariant recognition the visual system must be presented with images of an object at different locations on the retina. Position invariance can be learned by imposing temporal continuity on the response of a network to temporal sequences of patterns undergoing transformation ([2], [4], [5], and [10]). However, when the motion of the objects in the external world produces the required presentation of the object image across the retina, the difference in the appearance of a moving object is generally greater as the displacement increases. This may cause problems in some of the current position-invariant approaches. For example, [4] reported that the learning result was very sensitive to the time scale and the temporal structure in the input.

We introduce attention shifts as a key factor in the learning of position invariance. For the task of learning position invariance, the advantage of treating image feature displacements as being due to attention shifts is the fact that attention shifts are rapid, and that there is a neural command signal associated with them. The rapidity of the shift means that learning can be concentrated to take place only in the short time interval around the occurrence of the shift. This focusing of the learning solves the problems with time-varying scenery that plagued previous methods, such as those proposed by Einhäuser et al. [4] and Földiák [5].

In this paper we present a temporal learning scheme where knowledge of the attention shift command is used to gate the learning process. A temporal perceptual stability constraint is used to drive the output of the network towards remaining constant across temporal sequences of saccade motions and attention shifts. We implement a four-layer neural network model, and test it on both real images and simulated data consisting of various geometrical shapes undergoing transformations.

2   Learning Model

We refer to *local feature* as the visual information falling within the attention window, which includes either a whole object at low resolutions or parts of an object in fine details. Therefore, the learning of position invariance is achieved at two different levels: one is at a coarse level where position-invariant representation of local features is learned; and the other is at a fine level, where the position-invariant representation of an object as a whole with high resolution is learned by temporally correlating local features across attention shifts.

The overall model being proposed is composed of two sub-modules, as illustrated in Figure 1. One is the *attention control module*, which generates attention-shift signals according to a dynamically changing saliency map mechanism. The attention-shift signals are used to determine the timing for learning. The module obtains as input local feature images from the input raw retinal images via a position-changing attention window associated with an attention shift. The second sub-module is the *learning module*, which performs the learning of invariant neural representations across attention shifts

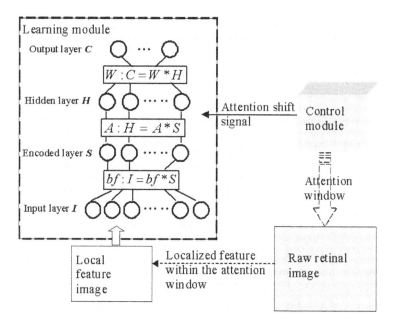

Fig. 1. System structure of the proposed neural network model

in temporal sequences. Two forms of learning, position invariant extraction of local features, and attention-shift invariant integration of object representation (an object is a composition of a set of local features), are triggered by the attention-shift signals. A temporal perceptual stability constraint is used to drive the output of the network towards remaining constant across the sequence of attention shifts.

2.1 Local Feature Selection Across Attention Shifts

Attention shift information is provided in our model by the *attention control module*. This module receives the retinal image as input. It constructs a saliency map mechanism [7] which is used to select the most salient area as the next attention-shift target. Currently our implementation uses grey-level images and we use orientation contrast and intensity contrast as saliency map features to form the saliency map. Intensity features, $I(\sigma)$, are obtained from an 8-level Gaussian pyramid computed from the raw input intensity, where the scale factor ranges from [0..8]. Local orientation information is obtained by convolution with oriented Gabor pyramids $O(\sigma, \theta)$, where [0..8] is the scale and [0°, 45°, 90°, and 135°] is the preferred orientation.

A Winner-Take-All algorithm determines the location of the most salient feature in the calculated saliency map to be the target of the next attention shift. An Inhibition-Of-Return (IOR) mechanism is added to prevent immediate attention shifts back to the current feature of interest, to allow other parts of the object to be explored. In the case of an overt attention shift, the target is foveated after the commanded saccade, and a new retinal image is formed. The new image is fed into the module as input for the next learning iteration. A covert attention shift, on the other hand, will not foveate the attended target, and therefore the subsequent retinal image input remains unchanged. Both overt and covert attention play an equal role in selecting local features, which are obtained when part of an object falls in the attention window before and after attention shifts. The resulting local features are fed into the input layer of the four-layer network.

2.2   Learning of Position-Invariant Local Features

The learning of position-invariant local features is based on the Clark-O'Regan approach ([3] [11]), which used a Temporal-Difference learning schema to learn the pair-wise association between pre- and post- motor visual input data, leading to a constant color percept across saccades. The learning takes place in the lower layers of the *learning module* from the input layer to the hidden layer. Our aim is to reduce the computational requirements of the Clark-O'Regan model while retaining the capability of learning position invariance. We make a modification to the learning rule, using temporal differences over longer time scales rather than just over pairs of successive time steps. In addition, we use a sparse coding approach to re-encode the simple neural responses, which reduces the size of the association weight matrix and therefore the computational complexity. The learning is triggered by the overt attention shift signal each time when a foveating saccade is committed.

We use a sparse coding approach [14] to reduce the statistical redundancies in the input layer responses. Let I denote the input layer neuronal responses. A set of basis functions bf and a set of corresponding sparsely distributed coefficients S are learned to represent I:

$$I(j) = \sum_i S_i * bf_i(j) \Rightarrow I = bf * S \tag{1}$$

The basis function learning process is a solution to a regularization problem that finds the minimum of a functional E. This functional measures the difference between the original neural responses I and the reconstructed responses $I' = bf * S$, subject to a constraint of sparse distribution on the coefficients S:

$$E(bf, S) = \frac{1}{2} \sum_j [I_j - \sum_i S_i * bf_{ij}]^2 + \alpha \sum_i Sparse(S_i) \tag{2}$$

where $Sparse(a) = ln(1 + a^2)$.

The sparsely distributed coefficients S become the output of the encoded layer. A weight matrix A between the encoded layer and the hidden layer serves to associate the encoded simple neuron responses related to the same physical stimulus at different retinal positions. Immediately after an attention shift takes place, this weight matrix is updated according to a temporal difference reinforcement-learning rule, to strengthen the weight connections between the neuronal responses to the pre-saccadic feature and that to the post-saccadic feature.

The position-invariant neuronal response in the hidden layer H is represented by the following equation:

$$H = A * S \tag{3}$$

The updating is done with the following temporal reinforcement-learning rule:

$$\Delta A(t) = \eta * \{[(1 - \kappa) * R(t) + \kappa * (\gamma * H(t) - \hat{H}(t-1))] * \hat{S}(t-1)\} \quad (4)$$

where

$$\Delta \hat{H}(t) = \alpha_1 * (H(t) - \hat{H}(t-1)) \quad (5)$$

$$\Delta \hat{S}(t) = \alpha_2 * (S(t) - \hat{S}(t-1)) \quad (6)$$

Here κ is a weighting parameter to balance the importance between the reinforcement reward and the temporal output difference between successive steps. The factor γ is adjusted to obtain desirable learning dynamics. The parameters η, α_1 and α_2 are learning rates with predefined constant values. The short-term memory traces, $\hat{H}$ and $\hat{S}$, of the neural responses in the hidden layer and the encoded layer, are maintained to emphasize the temporal influence of a response pattern at one time step on later time steps. The constraint of temporal perceptual stability requires that updating is necessary only when there exists a difference between current neural response and previous neural responses kept in the short-term memory trace. Therefore the learning rule incorporates a Hebbian term between the input trace and the output trace residuals (the difference between the current and the trace activity), as well as that between the input trace and the reinforcement signal.

Our approach is able to eliminate the limitations of Einhäuser *et al*'s model [4] without imposing an overly strong constraint on the temporal smoothness of the scene images. For example, in the case of recognizing a slowly moving object, temporal sampling results in the object appearing in different positions on the retina, which could cause temporal discontinuity in the Einhäuser *et al* model. But this will not affect the learning result of our approach because it employs a rapid overt attention shift to foveate the target object and only requires the temporal association between images before and after the attention shift.

2.3   Position-Invariant Representation of an Object Across Attention Shifts

Given that position-invariant representations of local features have been learned in the lower layers, an integration of local features of an object with fine details can be learned in the upper layers in the *learning module*. The representation of an object is learned in a temporal sequence as long as attention shifts stay within the range of the object. Here we assume that attention always stays on the same object during the recognition procedure of an object even in the presence of multiple objects. In our experiments this assumption was enforced by considering only scenes containing a single object.

A Winner-Take-All interaction ensures that only one neuron in the output layer wins the competition to actively respond to a certain input pattern. A fatigue process gradually decreases the fixation of interest on the same object after several attention shifts. This helps to explore new objects in the environment. For simplicity, in our experiments we restrict our scenes to contain only one object, we nonetheless implement the fatigue effect mechanism to simulate conditions such that an output layer neuron becomes fatigued after responding to the same object for a period of time. The fatigue effect is controlled by a Fixation-Of-Interest function $FOI(u)$. A value of u is kept for each output layer neuron, in an activation counter initialized to zero. Each counter traces the recent neural activities of its corresponding output layer neuron. The counter automatically increases by 1 if the corresponding neuron is activated, and decreases by 1 until 0 if not. If a neuron is continuously active over a certain period, the possibility of its next activation (i.e., its fixation of interest on the same stimulus) is gradually reduced, and thus allows other neurons to be activated. A Gaussian function of u^2 is used for this purpose:

$$FOI(u) = \exp(-u^4/\sigma^2) \tag{7}$$

An output layer neural response C_0 is obtained by multiplying the hidden layer neural responses H with the integration weight matrix W. C_0 is then adjusted by multiplying with $FOI(u)$, and is biased by the local estimation of the maximum output layer neural responses (weighted by a factor $\chi < 1$).

$$C' = C_0 * FOI(u) - \chi * \tilde{C}_0 \tag{8}$$

If C'_i exceeds a threshold, the corresponding output layer neuron is activated ($C_i = 1$).

The temporal integration of local features is accomplished by dynamically tuning the connection weight matrix between the hidden layer and the output layer. Neuronal responses to local features of the same object can be correlated by applying the temporal perceptual stability constraint that output layer neural responses remain constant over time. The constraint is achieved using a back-propagation term, where the short-term activity trace of the output neurons acts as a teaching credit to force the reduction of the temporal difference of neuronal responses between successive steps.

Given as input the position-invariant hidden layer neural responses H from the output of the lower layers, and as output the output layer neural responses C, the weight matrix W is dynamically tuned using the learning rule as follows:

$$\Delta W(t) = \gamma * [\lambda * \hat{C}(t) - (1 - \lambda) * (\hat{C}(t) - C(t))] * H(t) \tag{9}$$

where

$$\Delta \hat{C}(t) = \alpha * (C(t) - \hat{C}(t - 1)) \tag{10}$$

The short-term activity trace $\hat{C}$ acts as an estimate of the output layer neuron's recent responses. The first term of the learning rule emphasizes the Hebbian connection between the output neuronal activity trace and the current hidden layer neuronal response; and the second term is the back-propagation term, which drives the updating of the weights towards constant output. The factor λ balances the importance between the Hebbian connection and the temporal continuity constraint.

3   Simulation and Results

In our model, position invariance is achieved when a set of neurons can discriminate one stimulus from others across all positions. Furthermore, the neural response should remain constant while attention shifts over the same object. We refer to "a set of neurons", as our representation is in the form of a population code, in which more than one neuron may exhibit a strong response to one set of stimuli. Between each set of neurons there might be some overlap, but the combination of actively responding neurons are unique, and can therefore be distinguished from each other.

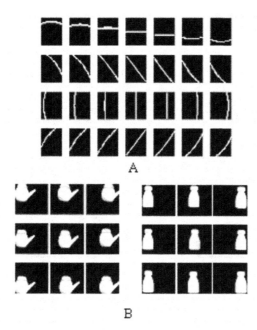

A

B

Fig. 2. Training data set of computer-simulated retinal images of lines (A) and image sequences of two real objects (B)

3.1   Demonstration of Position-Invariant Representation of Local Features

To demonstrate the process of position-invariant local feature extraction, we focus on the lower three layers: the input layer, the encoded layer and the hidden layer. The expected function of these layers is to produce constant neuronal responses to local features of an object at different positions after learning. We use two different test sets of local features as training data at this stage, a set of computer-generated simple features such as oriented lines at different position in a wiping sequence, and a set of computer-modified picture of real objects.

We first implemented a simplified model that has 648 input layer neurons, 25 encoded layer neurons and 25 hidden layer neurons for testing with the first training data set. The receptive fields of the input layer neurons are generated by Gabor functions over a 9x9 grid of spatial displacements each with 8 different orientations evenly distributed from 0 degree to 180 degree.

The first training image set is obtained by projecting straight lines of 4 different orientations ([0 °, 45 °, 90 °, and 135 °]) through a pinhole eye model onto 7 different positions of a spherical retinal surface. The simulated retinal images each have a size of 25x25 pixels. Figure 2A shows the training data set.

It was found in our experiment that some neurons in the hidden layer responded more actively to one of the stimuli regardless of its positions on the retina than to all other stimuli, as demonstrated in Figure 3. For example, neuron #8 exhibits a higher

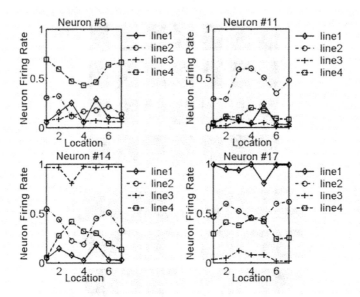

Fig. 3. Neural activities of the four most active hidden layer neurons responding to computer simulated data set at different positions

firing rate to line #4 than to any of the other lines, while neuron #17 responds to line #1 most actively. The other neurons remain inactive to the stimuli, which leave possible space to respond to other stimuli in the future.

It was next shown that the value of the weighting parameter κ in Equation 4 had a significant influence on this sub-module performance. To evaluate the performance, the standard deviation of activities of the hidden layer neurons are calculated when the sub-module is trained with different values of $\kappa(=0, 0.2, 0.5, 0.7$ and $1)$. The standard deviation of the neural activities is calculated over a set of input stimuli. The value stays low when the neuron tends to maintain a constant response to the temporal sequence of a feature appearing at different positions. Figure 4 shows the standard deviation of the firing rate of the 25 hidden layer neurons with different values of κ. The standard deviation becomes larger as κ increases. This result shows that the reinforcement reward plays an important role in the learning of position invariance. When κ is near 1, which means the learning depends fully on the temporal difference between stimuli before and after a saccade, the hidden layer neurons are more likely to have non-constant responses. Although there seems no much difference on the performance when the values of κ become very low, the reason why we keep the term of the temporal perceptual stability constraint in the learning rule is that this constraint forces the learning towards a constant state more quickly therefore makes the learning speedy. Choosing a proper value of κ is to balance the tradeoff between the performance and the speed. In practice we choose a non-zero but a rather small value of κ, for example $\kappa = 0.2$ for the most of the simulations we have run.

In our second simulation we tested image sequences of real world objects, such as a teapot and a bottle (Figure 2B). The images were taken by a digital camera with the

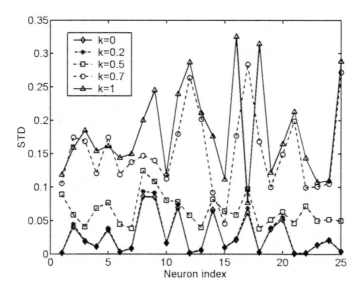

Fig. 4. Comparisons of position-invariant sub-module performance with varied weighting parameter κ ($\kappa = 0, 0.2, 0.5, 0.7, 1$), using a measurement of standard deviation of each neuronal response to a stimulus across different positions

center of its lens at nine different positions. Each image has a size of 64x48 pixels. The number of neurons in the encoded layer and the hidden layer has been increased from 25 to 64 from the numbers used in the previous experiment. This was required because the size of the basis function set to encode the sparse representations should also increase as the complexity of the input images increases.

Figure 5 shows the neural activities of the four most active neurons in the hidden layer when responding to the two image sequences of a teapot and a bottle respectively. Neurons #3 and #54 exhibit relatively strong responses to the teapot across all nine positions, while neuron #27 mainly responds to the bottle. We also have neuron #25 with strong overlapping neural activities to both stimuli. Satisfying our definition of position invariance, the sets of neurons that have relatively strong activities are different from each other.

3.2   Demonstration of Position-Invariant Representation of Objects

As the early learning process of integration is essentially random and has no effect on the later result, we use a gradually increasing parameter to adjust the learning rate of integration. This parameter can be thought of as an evaluation of the gained experience at the basic learning stage. The value of this parameter is set near 0 at the beginning of the learning, and near 1 after a certain amount of learning, at which point the position-invariant extraction process is deemed to have gained sufficient confidence in its experience on extracting position-invariant local features.

For simplicity, in this experiment we use binary images of basic geometrical shapes such as rectangles, triangles and ovals. These geometrical shapes are, as in the previous

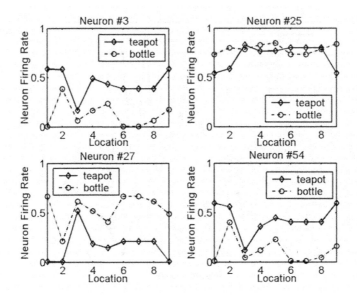

Fig. 5. Neural activities of the four most active hidden layer neurons responding to two real objects at different positions

experiment, projected onto the hemi-spherical retinal surface through a pinhole. Their positions relative to the fovea change as a result of saccadic movements.

Here we use an equal-weighted combination of intensity contrast and orientation contrast to compute the saliency map, as they are the most important and distinct attributes of the geometrical shapes we use in the training. A Winner-Take-All mechanism is

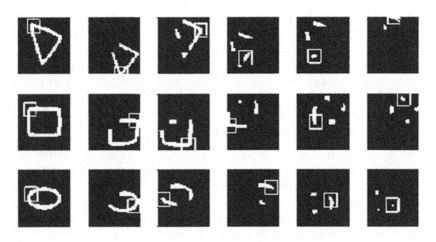

Fig. 6. Dynamically changing saliency maps for three geometrical shapes after the first 6 saccades following an overt attention shift. The small bright rectangle indicates an attention window centered at the most salient point in the saliency map

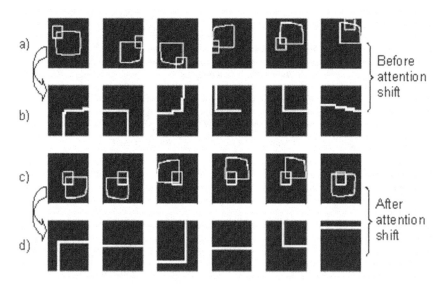

Fig. 7. Local features of a rectangular shape before (b) and after (d) an overt attention shift. a) and c) show retinal images of the same rectangle at different positions due to overt attention shifts

employed to select the most salient area as the next fixation target. After a saccade is performed to foveate the fixation target, saliency map is updated based on the newly formed retina, and a new training iteration begins. Figure 6 shows a sequence of saliency maps dynamically calculated from retinal images of geometrical shapes for a sequence of saccades. Local features of an object are obtained by falling in the attention window through these saliency maps as shown in Figure 7. Figure 7b and 7d show a sequence of pre- and post-saccadic local features of the retinal images of a rectangular shape falling in a 25x25 pixel attention window respectively.

Position-invariant representations of these local features are achieved from the previous learning steps in the lower layers, and they are fed into the integration procedure for the learning of invariance across attention shifts.

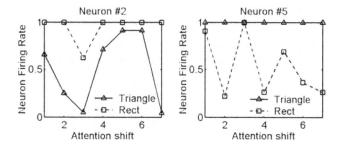

Fig. 8. Neural activities of the two most active output layer neurons responding to two geometric shapes across attention shifts

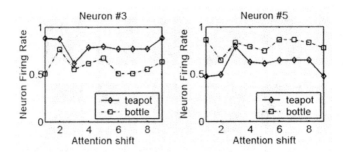

Fig. 9. Neural activities of the two most active output layer neurons responding to two objects (a teapot and a bottle) across attention shifts

We show in Figure 8 some of the output layer neural responses (neuron #2 and neuron #5) to two geometrical shapes: a rectangle and a triangle. Neuron #2 responds to the rectangle more actively than to the triangle, while neuron #5 has higher firing rate to the triangle than to the rectangle. Figure 9 shows another experiment result on two real objects, where output layer neuron #3 and neuron #5 have stronger response to a teapot and a bottle respectively.

4   Discussions

Our proposed approach is different from the dynamic routing circuit of Olshausen et al. [13], which forms position- and scale- invariant representations of objects for recognition via attention shifts. Firstly, attention shift plays different roles in both models. In Olshausen's model attention shift is merely for feature segmentation and selection. In our approach, attention shift also provides a reinforcement signal for learning, and there is a constraint on the learning rule that the output should remain constant when attention stays on the object. Secondly, Olshausen's model does not take into account image distortions due to the non-uniformity of the retina and the nonlinearity of projection onto the hemi-spherical retina when foveating eye movements take place. The object to be recognized undergoes simply translation and rescaling and therefore having little shape dissimilarity at each position and scale, while in practice distortions might lead to a greatly different appearance when the object is projected on the periphery. Our model is designed to consider the position-related distortions in the retinal input, leading to a position-invariant representation of an object. The overt attention shifts employed in our model brings an object in the periphery into the fovea, and correlates the before- and after- saccade retinal information to form a canonical representation regardless of its distortion.

The current approach is trained on one object per image. A further work has to be done on multiple objects in the scene with different background. To keep attention mostly staying on the same object during the learning, we can modify the formation of the saliency map by introducing a top-down stream which has preference to the neighborhood of the current focus of attention. It is true that there can be attention shifts that cross between objects. In practice, however, a given object will appear in conjunction with a wide range of different objects from time to time. Thus, during learning in a complex uncontrolled environment, there will be few persistent associations to be made across inter-object attention shifts. For intra-object attentions shifts, on the other hand, the features will, in general, be persistent, and strong associations can be formed. In our view (which we have not tested) is that inter-object attention shifts will only slow down learning and not prevent it. The other methods for position-invariance also have problems with multiple objects, and they assume some form of scene segmentation to have been done.

The work described here mainly concerns learning of a constant neural representation of objects with respect to the position variance and the corresponding position-related distortions. The representation can be further applied to the task of object recognition using some supervised learning rule to associate the neural coding of an object with a certain object class. While recognizing an object across a sequence of attention shifts, the activities of the output neurons can be viewed as an evaluation of the confidence on a certain object. Although during the attention shifts, only parts of the object can be focused and some of the parts might coincident with other objects, we hypothesize that for most objects the set of their salient points are different. Therefore a temporal trace of the output neural activities over attention shifts is likely to represent an object as a whole uniquely in the form of its neural coding and can distinguish one from the others.

5   Conclusions

In this paper we have presented a neural network model that achieves position invariance incorporated with visual-related self-action signals such as attention shifts. Attention shifts play an important role in the learning of position invariance. Firstly, attention shifts are the primary reason for images of object features to be projected at various locations on the retina, which enables the learning focus on position invariance. Secondly, attention shifts actively select local features as input to the learning model. Thirdly, the motor signals of attention shift are used to gate the learning procedure.

We implemented a simplified version of our model and tested it with both computer-simulated data and computer-modified images of real objects. In these tests local features were obtained from retinal images falling in the attention window by an attention shift mechanism. The results show that our model works well in achieving both position invariance, regardless of retinal distortions.

References

1. Bandera, C., Vico, F., Bravo, J., Harmon, M., and Baird, L., Residual Q-learning applied to visual attention, *Proceedings of the 13th International Conference on Machine Learning*, (1996) 20–27
2. Becker, S., Implicit learning in 3D object recognition, The importance of temporal context. *Neural Computation* 11(2), (1999) 347–374
3. Clark, J.J. and O'Regan, J.K., A Temporal-difference learning model for perceptual stability in color vision, *Proceedings of 15th International Conference on Pattern Recognition* 2, (2000) 503–506

4. Einhäuser, W., Kayser, C., König, P., and Körding, K. P., Learning the invariance properties of complex cells from their responses to natural stimuli, *European Journal of Neuroscience* **15**, (2002) 475–486

5. Földiák , P., Learning invariance from transformation sequences, *Neural Computation* **3** (1991) 194–200

6. Henderson, J.M., Williams, C.C., Castelhano, M.S., and Falk, R.J., Eye movements and picture processing during recognition, *Perception and Psychophysics* **65(5)**, (2003) 725–734

7. Itti, L., Koch, C. and Niebur, E., A model of saliency-based visual attention for rapid scene analysis, *IEEE Transactions on Pattern Analysis and Machine Intelligence* **20(11)**, (1998) 1254–1259

8. Kikuchi, M., and Fukushima, K., Invariant pattern recognition with eye movement: A neural network model, *Neurocomputing* **38-40**, (2001) 1359–1365

9. Koch, C., and Ullman, S., Shifts in selective visual attention: Towards the underlying neural circuitry. *Human Neurobiology* **4**, (1985) 219–227

10. Körding, K. P. and König, P., Neurons with two sites of synaptic integration learn invariant representations, *Neural Computation* **13**, (2001) 2823–2849

11. Li, M. and Clark, J.J., Sensorimotor learning and the development of position invariance, poster presentation at *the 2002 Neural Information and Coding Workshop*, Les Houches, France, (2002)

12. Minut, S., and Mahadevan, S., A reinforcement learning model of selective visual attention, *AGENTS2001*, (2001) 457–464

13. Olshausen, B.A., Anderson C.H., Van Essen D.C., A neurobiological model of visual attention and invariant pattern recognition based on dynamic routing of information, *The Journal of Neuroscience* **13(11)**, (1993) 4700–4719

14. Olshausen, B. A. and Field, D. J., Sparse coding with an overcomplete basis set: A strategy employed by V1?, *Vision Research* **37**, (1997) 3311–3325

Combining Conspicuity Maps for hROIs Prediction

Claudio M. Privitera[1], Orazio Gallo[2], Giorgio Grimoldi[2],
Toyomi Fujita[1], and Lawrence W. Stark[1]

[1] Neurology and Telerobotics Units, Optometry School,
University of California, Berkeley, 94720, CA
[2] Department of Bioengineering of Politecnico di Milano,
P.zza Leonardo da Vinci, 20133 Milan

Abstract. Bottom-up cortical representations of visual conspicuity interact with top-down internal cognitive models of the external world to control eye movements, EMs, and the closely linked attention-shift mechanisms; to thus achieve visual recognition. Conspicuity operators implemented with image processing algorithms, IPAs, can discriminate human Regions-of-Interest, hROIs, the loci of eye fixations, from the rest of the visual stimulus that is not visited during the EM process. This discrimination generates predictability of the hROIs. Further, a combination of IPA-generated conspicuity maps can be used to achieve improved performance over each of the individual composing maps in terms of hROI predictions.

1   Introduction

Saccadic eye movements and the closely linked mechanisms of visual attention shifts have been attracting scientists for decades. These are very complex functions which utilize many different aspects of the human nervous system from early visual areas and brain stem motor competencies up to high level semantic elaborations. The classical work performed by Yarbus back in the sixties on task- and intention-dependent eye movements, EMs, [1], is still acknowledged and very influential.

Several models have been presented in the neuroscience and computational vision community. In general, these models seem to support theories by which selective attention shifts are strictly related to, or even controlled by a visual multi-feature conspicuity cortical representation; see, for example the pioneering work of Koch and Ullman, [2] and more recently an interesting review by Itti and Koch, [3] and the approach of Parkhurst et. al [4]. Visual attention, which is linked to EMs, is thought to be a process guided by this map (likely located in the frontal eye field, [5], although neurophysiological data support the existence of several conspicuity maps in different cortical areas such as temporal and pre-frontal cortex, [6], [7]). Such a map is also thought to preserve the topography of the external viewed world and measures, in the z-coordinate, levels of conspicuity or saliency; the highest points or local maxima of the map define the loci (here referred to as hROIs, human Regions-of-Interest) of sequential steps or saccades that are governed by a sequential winner-take-all procedure [2], [3].

The nature and number of cortical topographical conspicuity representations vary; in general they must be related to bottom-up early visual conspicuity features such as

L. Paletta et al. (Eds.): WAPCV 2004, LNCS 3368, pp. 71–82, 2005.

center-surrounded operators, high-contrast contours and areas, and color and selective orientations [8], [9], [3]. The work by Reinagel and Zador, [10] for example, shows how subjects look at hROIs that have high spatial contrast and poor pair-wise pixel correlation; a definition of internal intensity non-uniformity. Similarly, higher-order statistics have recently been used by Krieger et al. [11] to characterize hROIs and to propose features like corners, occlusions, and curved lines as salient detectors. Different representations and different visual scales can be processed at different sequential temporal stages during active looking, [12] or in parallel and then weighted and combined into a final *supra-dimensional* or *supra-feature* conspicuity map [13].

Cognition must influence these conspicuity representations in a top-down manner, [14], however, how this influence is exercised and integrated in the overall control of EMs/attentional shifts is still not well understood and open to different interpretations. Interesting models wherein top-down and bottom-up processing are integrated into Bayesian or Markovian frameworks can also be found in the literature [15], [16], [17].

Our standpoint is strongly based on the scanpath theory put forward by Noton and Stark, [18], [19]. The scanpath was defined on the basis of experimental findings. It consists of sequence of alternating saccades and fixations that repeat themselves when a subject is viewing a picture. Only ten percent of the duration of the scanpaths is taken up by the collective duration of the saccadic eye movements, which thus provide an efficient mechanism for traveling over the scene or regions of interest. Thus, the intervening fixations or foveations onto hROIs have at hand ninety percent of the total viewing period. Scanpath sequences appear spontaneously without special instructions to subjects and were discovered to be repetitive, [18], [19].

The scanpath theory proposes that an internal spatial cognitive model controls both perception and the active looking EMs of the scanpaths sequence, [18], [19], and further evidence for this came from new quantitative methods, experiments with ambiguous figures, [20] and more recently from experiments on visual imagery [21], [22] and from MRI studies on cooperating human subjects [23].

A top-down internal cognitive model of the external world must necessarily control not only our recognition but also the sequence of EM jumps which direct the high-resolution and centrally located fovea into a sequence of human regions-of-interest, hROI loci. We usually refer to these hROI loci as *informative* regions because they carry the information needed (and thus searched) by our cognition to validate or propose new internal cognitive models during active looking EM scanpaths, [24]. Recognition is re-cognition, which signifies to know again, to recall to mind a predefined model; eye movements and cognition must act in synergy.

Conspicuity and informativeness are thus interconnected, and conspicuity can be used to predict informativeness. Our studies have indeed demonstrated that bottom-up image processing algorithms, IPAs, can be successfully used, together with an eccentricity clustering procedure to generate algorithmic Regions-of-Interest, aROIs. These aROIs are able to predict the hROI loci of human observers, obtained during general viewing EM experiments, [25]. The level of the prediction, meaning the proximity of an IPA-generated sequence of aROI loci and an experimental instance of an hROIs sequence, or scanpath, can be measured by a spatial metric, S_p, which has been defined on the basis of physiological and experimental observations. The metric S_p is important in many applications, [26].

Different IPAs yield different results depending on the class of images and the specific task involved, [27]. Also, as discussed above, different IPAs can be combined together into a *supra-feature* IPA. The resulting combination might be used to achieve improved performance over each of the individual IPAs in terms of hROIs S_p-prediction and in terms of consistency for larger numbers of images. A study specifically designed for the geological/planetary exploration application, [28], has already shown interesting preliminary results.

In the present paper we want to elucidate the important relationship between image conspicuity distribution and experimental human scanpath hROIs, for different image classes and for more general viewing conditions; the S_p metric will be utilized for this purpose. Heterogeneous sets of bottom-up features can be combined with ad-hoc selection and fusion processes. The results will confirm that such a combination can perform better than using only a single feature.

2   Sp Similarity Metric and Saliency Discrimination

2.1   Experimental hROI Database

Standard procedures were used to extract hROIs, the loci of eye fixations, ([22], circles superimposed on the raw eye movement data, Figure 1, upper left panel); an average of seven hROIs was collected per image, which usually corresponds to about three seconds of viewing. Different classes of images for a total of forty images were used; they were photos of *Interior*, i.e., environments like a theater, *Traffic*, i.e., road intersection in a big city, *Natural*, i.e., a canyon or a lake, *Geological*, i.e., rocks or terrain, *Ambient*, i.e., a kitchen or a bedroom, *Urban*, i.e., snapshots of city life.

Data were from new EM experiments and also provided by the Neurology and Telerobotics Units archive. The eye tracker system used in the Units resides in a Pentium3 two-processor computer which controls both visual stimuli presentation and the video camera analysis. The first corneal Purkinje reflection from an infrared light source is tracked in real time using a flying-spot algorithm and then later processed by a parsing algorithm for the identification of hROIs. For each image, two calibration sessions composed by a sequence of 3×3 fixation points are used to map video camera eye positions into the stimulus linear space. The experiment is repeated if calibration drift or similar deleterious experimental artifacts are detected during or between the two calibrations. The subjects were secured in a head-rest optometric apparatus and no specific viewing instructions were provided.

2.2   Algorithmic Identification of Regions-of-Interest, aROIs

Several sources supplied the collection of IPAs; they were in general inspired by early-stage neurophysiology, such as center-surround filters, localized projectors, or concentric detectors; other IPAs were simply based on intuition or different approaches presented in the literature. Sixteen IPAs were extracted from the collection and used in this study. They were all defined on a local support size corresponding to two degrees of the visual

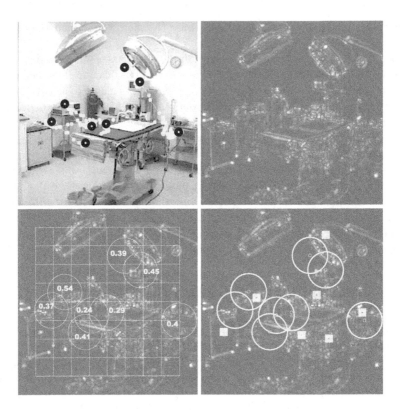

Fig. 1. Loci of eye fixations, hROIs (black small circles, upper-left panel), are extracted from raw eye movement data (white trace, upper-left panel) using standard procedures. An energy map is created using a symmetry operator (see text, upper-right panel). Conspicuity is measured for the hROIs (large circles, bottom-left panel, peak of energy values are also reported) and for the rest of the image (grid, bottom-left panel). The energy map is also used to identify algorithmic Regions-of-Interest, aROIs (squares, bottom-right panel) whose loci can finally be compared with hROIs

field, similar to the subtended angle of the fovea in our experimental setup. Algorithms A_1 to A_4 are four different Gabor filters rotated by $0, 45, 90, 135$ degrees, [29]. Algorithm A_5 was a difference in the gray-level orientation operator, [30]. We then have A_6, local entropy, [25], A_7, Laplacian of a Gaussian, [31], A_8, contrast operator, [25], A_9, x-like operator, [25], A_{10}, radial symmetry, [32], A_{11}, quasi receptive, [25], $A_{12} - A_{14}$, frequency analysis, low, medium, high pass filters based on DCT, [25], $A_{15} - A_{16}$, color opponence operators, red-green and yellow-blue, [30].

Each IPA gives emphasis to a different bottom-up visual conspicuity when applied to an image and correspondingly defines its own energy map. A symmetry operator, for example, highlights symmetry, and the resulting output energy map (Figure 1, upper right panel) defines both the level of this feature (the z-dimension is coded in grey-level, where brighter patches represent centers of high symmetricity) and its spatial distribution over the image. Local maxima in the output energy map define peaks of conspicuity and the corresponding loci are possible candidates to be selected as the algorithmic Regions of Interest, aROIs.

An eccentricity clustering procedure was introduced in [25] to select eccentric-located local maxima as the final aROI loci; the algorithm was based on an incremental merging of neighboring local maxima, and the highest peak locus was finally held for each local neighborhood. There is probably little biological plausibility for the mechanism but important practical convenience; the selection of aROI maxima is a parameter-free procedure independent of scale and image size. The eccentricity is necessary to maximize the extent of the image that is covered with a limited number of hROIs, EM jumps. The winner-take-all algorithm utilized by many authors (see for example [3]) in conjunction with an inhibition-of-return (or *forgetting function*) mechanism, [33] has more neuronal justifications; of course, the spatial size of the inhibition needs to be properly set in this case. We decided on the same size defined above for the IPA local support and corresponding to two degrees of the visual field. In the winner-take-all algorithm absolute local maxima are iteratively selected as aROIs; for each absolute maximum, the forgetting function zeroes all the energy around it and the aROI selection process continues with the next absolute maximum. The top seven aROIs were retained for each image. How these two selection procedures affect the final S_p outcome is still under investigation; preliminary results, however, seem to indicate that the final distribution of aROIs is independent of the selection procedure (see also [34]).

2.3   Sp Similarity Index

Each image and experimental instance is thus represented by a pair of vectors; the vector of hROIs (Figure 1, lower left panel, circles) is the original scanpath and is experimentally defined for each single subject and experimental trial. The vector of aROIs (Figure 1, lower right panel, squares) is the artificial scanpath and is computed from the IPAs as just described. The loci similarity between hROIs and aROIs (but also between two hROIs or two aROIs) is expressed by a spatial similarity index, S_p, [25] which represents the percent of experimental hROIs which are in close proximity to at least one aROI. An S_p value of 1 means that all hROIs are captured within the aROIs; zero means a complete spatial dissociation between the two vectors. The threshold distance defining this proximity is approximately 2 degrees (Figure 1, lower two panels, big circumferences) and is based on experimental intra-subject EM observation (for a full review see discussions on the EM Repetitive similarity in [25]); for example, two aROIs are in two (out of eight) different hROI circumferences (Figure 1, lower right panel), which yields in this case an S_p value equal to $2/8 = 0.25$.

Randomly generated aROIs can fall by chance within an hROI; this explain why the S_p similarity of such randomly generated aROIs with human scanpath, averaged for all images and repetitions, is 0.2, what we considered a bottom anchor or the lowest fiduciary point for S_p. The opposite (top) anchor or highest fiduciary point is the inter-subject S_p similarity whose average is about 0.6; this is the average of the scanpath S_p similarity indices when different subjects look at the same picture (then averaged for all pictures). A value of 0.6 says that an average of 60 percent of their hROIs cohere which indicates that viewers were fairly consistent in identifying regions of interest. This is an important results for this study, in fact, algorithms cannot be expected to predict human fixations better than the coherence among fixations of different persons.

3   Saliency Discrimination of Eye Fixation Loci

In human scanpaths, only a small portion of the entire image falls on the fovea, the high-resolution area of the retina, to be sequentially attended and analyzed by the brain; the rest of the image is processed only at much lower resolution by peripheral vision and is ignored by the eye fixation sequence. Subjects tend to repeat their scanpaths, and also cohere with other subjects in the loci of fixations, when they look at the same image (see above, [25] and the recent review in [22]).

This visual image spatial discrimination, which is implicit in the EM process, can be also studied in terms of the conspicuity distribution of a given IPA operator; how different is the distribution within the hROIs compared with the rest of the image that was not selected by the experimental scanpath? A measure that can be carried out by the D-prime index from the theory of signal detection:

$$D' = \frac{\mu_{max(hROIs)} - \mu_{max(NhROIs)}}{\sqrt{\sigma^2_{max(hROIs)} + \sigma^2_{max(NhROIs)}}} \tag{1}$$

In the D-prime definition above, $max(hROIs)$ is the peak of the energy in each hROI circumference (Figure 1, lower-left panel). The rest of the image is divided in a regular grid of 2×2 degree NhROI (an acronym for Non-hROI) blocks and for each of these blocks, the corresponding energy peak is also retained and used in the equation ($max(NhROIs)$). The contour of the image is very unlikely to be fixated during human scanpath and is thus not considered in the D-prime definition (Figure 1, lower-left panel). A positive value for D-prime indicates that the IPA energy (or more specifically the peak energies) inside the hROIs is on the average greater than that for the rest of the image – a necessary condition for a given IPA to be chosen as a conspicuity model for that specific hROI set.

Values of D-prime and S_p are indeed well correlated, (Figure 2, a correlation of 0.73); each $< S_p, D - prime >$ data point corresponds to a unique ternary: one hROIs instance, one image and an IPA. High values of D-prime conspicuity discrimination correspond to optimal scanpath S_p predictability; values of S_p that are low or below the random bottom anchor level are associated with poor or even negative discriminability. This indicates a spatial mismatch of the energy distribution with the hROIs portion of the image having less energy than the outside portion. Positive values of D-prime correspond to S_p similarities that are above the random bottom anchor $S_p = 0.2$. Note that most of the $< S_p, D - prime >$ curve (Figure 2) lays on the second part of the plot, above zero, which indicates that a positive discrimination is present in most of the experimental hROIs instances.

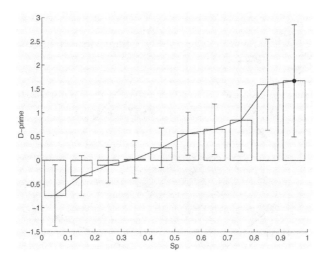

Fig. 2. Correlation between D-prime conspicuity discriminability of hROIs and the corresponding S_p hROI-aROI loci similarity. Positive values of D-prime correspond to S_p similarities that are above the random bottom fiduciary level, $S_p = 0.2$

4   Polynomial Combination of Energy Maps

Different energy maps can be weighted and combined together in a polynomial manner into a *supra-feature* map which is hereafter referred to as A* (Figure 3). We want to derive A* and verify that it can in fact improve the spatial S_p matching of aROIs with the experimental hROIs.

4.1   Optimization for Each Single Experimental hROIs Instance

We initially tried to define the best combination for each single image and hROIs; the objective was to determine the vector of weights, $\overline{\alpha}$, which maximizes the corresponding S_p for A*, a polynomial of the input energy maps $A_1, A_2, ..., A_{16}$ (Figure 3). An optimization problem that we implemented in Matlab with the built-in medium scale algorithm of the constrained minimization routine (based on the Sequential Quadratic Programming method, [35]).

Although S_p is the index to be maximized, D-prime is a more suitable objective function for the optimization; it is a continuous n-dimensional function, appropriate for a canonical gradient-descent searching process in the $\overline{\alpha}$ space. The metric S_p is discontinuous; an infinitesimal variation in the input space $\overline{\alpha}$ can cause a different energy maximum (aROI) to be selected in the A* map (as a consequence of the iterative absolute maximum selection procedure defined in the previous Section), which can result in a substantial change of S_p. Recall that S_p and D-prime are correlated (as shown in Figure 2). The object function is thus based on D-prime and defined as follow:

$$\max_{\overline{\alpha}}\{D'(A^*), \alpha_i > 0, \sum_i \alpha_i = 1\} \tag{2}$$

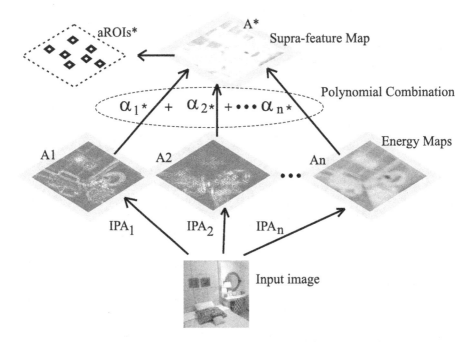

Fig. 3. A polynomial (weighted) combination of different conspicuity maps defines a *supra-dimension* conspicuity map A*

An image from the Ambient image can exemplify the optimization process (Figure 4); all IPAs were initially applied to the image and the best result was achieved with A_{15} (Figure 4, left) which yielded an S_p value of $5/7 = 0.71$. The aROIs* from the A* map (Figure 4, right) defined by the optimal set of weights $\overline{\alpha}*$ as resulted at the end of the optimization process, have an S_p value equal to 1. Each experimental hROIs data was used in a similar optimization process and the resulting combinatorial A* used to select aROIs* which could be finally compared with the corresponding hROIs. In general, more than 80% of the optimizations resulted in an A*-S_p that was higher or at least equal to the best of the single IPAs. The residual 20% of unsuccessful optimizations might be related to local maxima (not easily overcome in such a complex multi-dimensional search space) and to the not unitary correlation between the objective function D-prime and S_p.

4.2   Optimization for Image Classes

These single optimizations show that it is in general possible to define an improved polynomial combination of energy maps whose matching with experimental hROIs is superior to each of the composing single maps.

A more fundamental problem is to define a super polynomial combination A** which might be applied more consistently to all the images (and hROIs) of the same class. To define the corresponding vector of weights, $\overline{\alpha}**$, we tried initially to review the definition of the objective function in order to contemplate a multi-dimensional D-prime index

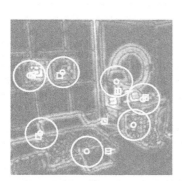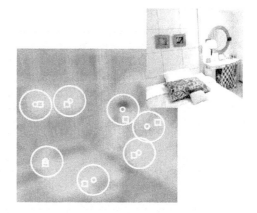

Fig. 4. Example of result for a single image/hROIs instance optimization; the best algorithm (algorithm $A_1 5$, left panel) compared with the final polynomial combination A* (right panel). Circles identify hROIs, squares aROIs; the A* S_p index is 7/7 vs. the 5/7 for the best algorithm

Table 1. The A** optimization for each single image class. The average S_p value of A** (right column) is always superior to the mean of all IPAs (middle column) and to the best IPA of each class (left column, the index (1 to 15) of the IPA is also reported in parentheses)

Class	Best IPA	Mean of IPAs	A**
Interior	(5) 0.54 (0.04)	0.45 (0.05)	0.62 (0.04)
Traffic	(9) 0.44 (0.09)	0.36 (0.05)	0.45 (0.04)
Natural	(6) 0.54 (0.07)	0.44 (0.06)	0.58 (0.05)
Geological	(4) 0.55 (0.06)	0.46 (0.06)	0.63 (0.06)
Ambient	(6) 0.6 (0.03)	0.49 (0.04)	0.64 (0.04)
Urban	(9) 0.57 (0.05)	0.39 (0.04)	0.59 (0.02)

for all the images in the data set and then to run the same optimization process; the computational repercussions were however too costly and thus impracticable. We then tried to use the same vectors, $\overline{\alpha}*$, already calculated from all the single optimizations. All classes of images were considered separately: for each class, all the vectors $\overline{\alpha}*$ were clustered using a K-means algorithm and the center of the densest was cluster retained as the final $\overline{\alpha}**$. This center in the multi-feature $\overline{\alpha}$ space likely represents the most common characteristics of that class and it is reasonable to select it as the corresponding representative. Other clustering possibilities are under investigation and evaluated with a clustering analysis procedure (such as the Akaike's criterion) that can serve to explicate the significance of each $\overline{\alpha}**$.

The results seem to support the initial hypothesis; the S_p behavior of the polynomial algorithm A**, averaged for all images and hROIs within a class, is consistently superior (see S_p value, table 1, right column) if compared with the mean of all composing IPAs (table 1, middle column) and with the mean of the best IPA for each class (table 1, left column). Almost all the A** (except for the Traffic class) yield an average S_p (table 1, right column) which is very close to the inter-subject S_p top anchor similarity of 0.6 discussed in the previous Section.

The best algorithm (table 1, number in parenthesis, left column) is different for each image class; also different classes generated different distributions of weights, $\overline{\alpha}^{**}$, (Figure 5, the values in the ordinate indicate the weight given to the corresponding IPA, in that specific class) further supporting the notion that different image types might be characterized by different types of conspicuity.

5   Conclusions

Different conclusions emerge from this study. First, experimental EM fixations, which are distributed over only a specific portion of the image, are discriminable in term of conspicuity and D-prime. Fifteen different IPAs were used to represent different type of conspicuities. More importantly, the level of discriminability correlates with the capability of a particular IPA to be used in an aROIs generation process, to match the human hROIs scanpath as expressed by the similarity index S_p. Secondly, a polynomial combination of IPAs yields to a better S_p prediction of experimental hROIs. This was demonstrated for most of the single experimental instances and for the complete set of images within a class. A procedure was proposed to define for each class, the vector of weights $\overline{\alpha}^{**}$ which defines the influence of each composing IPA in the *supra-feature* map A**.

As mentioned above, it has been known for some time that the implicit or explicit task-setting in which the subject is immersed can strongly modify the scanpath. Thus as a subject continually looks at the scene she may change her point of view, think

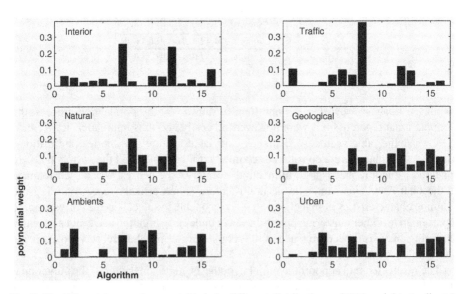

Fig. 5. Six different image classes yielded six different distributions of $\overline{\alpha}^{**}$ weights (ordinate) which define A**; algorithm index is in the abscissa

of different task-goals and modify the scanpath [1] (a planetary exploration task was recently investigated by Privitera and Stark, [28]). The vector of weights $\overline{\alpha}^{**}$ can thus be susceptible to the task-setting.

A discrimination of hROIs in term of visual conspicuity is experimentally plausible in human vision and analyzed in the literature from different perspectives; we propose an adaptive and composite nature for conspicuity which can vary for different image classes and probably tasks and applications. Cognition might modulate the combination of different types of conspicuity as a function of the internal TD hypothesis or the visual expectation: different environmental setting, for example Urban vs. Natural, can be associated to different clusters of conspicuity weights, $\overline{\alpha}^{**}$, that can be then utilized in the generation of the scanpath motor sequence.

Acknowledgements

We wish to thank Profs. Giancarlo Ferrigno, Alessandra Pedrocchi and Antonio Pedotti, all from the Department of Bioengineering of Politecnico di Milano, for their support and stimulating discussions. Our colleague in the laboratory, Dr. Yeuk Ho, was indispensable for his management of the laboratory computational resources and the eye-tracker system. A special thanks goes also to Dr. Tina Choi for her editorial assistance.

References

1. Yarbus, A.: Eye movements and vision. Plenum, New York (1967)
2. Koch, C., Ullman, S.: Shifts in selective visual attention: towards the underlying neural circuity . Human Neurobiology **4** (1985) 219–227
3. Itti, L., Koch, C.: Computational modelling of visual attention. Nature Neuroscience Reviews **2** (2001) 194–203
4. Parkhurst, D., Law, K., Niebur, E.: Modeling the role of salience in the allocation of overt visual attention. Vision Research **42** (2002) 107–123
5. Bichot, N.P., Schall, J.D.: Saccade target selection in macaque during feature and conjunction visual search. Visual Neuroscience **16** (1999) 81–89
6. Stein, J.E.: The representation of egocentric space in the posterior parietal cortex. Behavioral and Brain Sciences **15** (1992) 691–700
7. Nothdurft, H.: Salience from feature contrast: additivity across dimensions. Vision Researh **40** (2000) 1183–1201
8. Engel, F.L.: Visual conspicuity, visual search and fixations tendencies of the eye. Vision Research **17** (1977) 95–108
9. Findlay, J.M., Walker, R.: A model of saccade generation based on parallel processing and competitive inhibition. Behavioral and Brain Sciences **22** (1999) 661–721
10. Reinagel, P., Zador, A.M.: Natural scene statistic at the center of gaze. Network: Comput. Neural Syst. **10** (1999) 341–350
11. Krieger, G., Rentschler, I., Hauske, G., Schill, K., Zetzsche, C.: Object and scene analysis by saccadic eye-movements: an investigation with higher-order statistics. Spatial Vision **13** (2000) 201–214
12. Neri, P., Heeger, D.: Spatiotemporal mechanisms for detecting and identifying image features in human vision . Nature neuroscience **5** (2002) 812 – 816

13. Muller, H.J., Heller, D., Ziegler, J.: Visual search for singleton feature targets within and across feature dimensions. Perception and Psychophysics **57** (1995) 1–17
14. Henderson, J.M., Hollingworth, A.: in Eye guidance in reading and scene perception. In: Eye movements during scene viewing: an overview. North-Holland/Elsevier (1998)
15. Chernyak, D.A., Stark, L.W.: Top-down guided eye movements. IEEE Trans. on SMC, part B **31** (2001) 514–522
16. Rimey, R.D., Brown, C.M.: Controlling eye movements with hidden markov models. International Journal of Computer Vision **7** (1991) 47–65
17. Schill, K., Umkehrer, E., Beinlich, S., Krieger, G., Zetzsche, C.: Scene analysis with saccadic eye movements: top-downand bottom-up modeling. Journal of Electronic Imaging **10** (2001) 152–160
18. Noton, D., Stark, L.W.: Eye Movements and visual Perception. Scientific American **224** (1971) 34–43
19. Noton, D., Stark, L.W.: Scanpaths in Eye Movements during Pattern Perception. Science **171** (1971) 308–311
20. Stark, L.W., Ellis, S.R.: in Eye Movement: Cognition and Visual Perception. In: Scanpaths revised: cognitive models direct active looking. Lawrence Erlbaum Associates, Hillside, NJ (1981) 193–226
21. Brandt, S.A., Stark, L.W.: Spontaneous eye movements during visual imagery reflect the content of the visual scene. J. Cognitive Neuroscience **9** (1997) 27–38
22. Stark, L.W., Privitera, C.M., Yang, H., Azzariti, M., Ho, Y.F., Blackmon, T., Chernyak, D.: Representation of human vision in the brain: How does human perception recognize images? Journal of Electronic Imaging **10** (2001) 123–151
23. Kosslyn, S.: Image and Brain: The Resolution of the Imagery Debate. MIT Perss, Cambridge, MA (1994)
24. Stark, L.W., Choi, Y.S.: in Visual Attention and Cognition. In: Experimental Metaphysics: The Scanpath as an Epistemological Mechanism. Elsevier Science B.V. (1996) 3–69
25. Privitera, C.M., Stark, L.W.: Algorithms for Defining Visual Regions-of-Interest: Comparison with Eye Fixations. IEEE Trans. PAMI **22** (2000) 970–982
26. Privitera, C.M., Stark, L.W., Ho, Y.F., Weinberger, A., Azzariti, M., Siminou, K.: Vision theory guiding web communication. In: Proc. SPIE - Invited paper. Volume 4311., San Jose, CA (2001) 53–62
27. Privitera, C.M., Azzariti, M., Stark, L.W.: Locating regions-of-interest for the Mars Rover expedition. International Journal of Remote Sensing **21** (2000) 3327–3347
28. Privitera, C.M., Stark, L.W.: Human-vision-based selection of image processing algorithms for planetary exploration . IEEE Trans. Image Processing **12** (2003) 917–923
29. Daugman, J.G.: Two-dimensional spectral analysis of cortical receptive field profiles. Vision Research **20** (1980) 847–856
30. Itti, L., Kock, C., Niebur, E.: A model of saliency-based visual attention for rapid scene analysis. IEEE Trans. PAMI **20** (1998) 1254–1259
31. Marr, D., Hildreth, E.: Theory of edge detection. Proc. Roy. Soc. London **B207** (198) 187–217
32. Loy, G., Zelinsky, A.: Fast radial symmetry for detecting points of interest. IEEE Trans. PAMI **25** (2003) 959–973
33. Klen, R.M.: Inhibition of return . Trends Cogn. Sci **4** (2000) 138–147
34. Privitera, C.M., Krishnan, N., Stark, L.W.: Clustering algorithms to obtain regions of interest: a comparative study. In: Proc. SPIE. Volume 3959., San Jose, CA (2000) 634–643
35. Gill, P.E., Muray, W., Wright, M.H.: Practical Optimization. Academic Press, London (1981)

Human Gaze Control in Real World Search

Daniel A. Gajewski[1,4], Aaron M. Pearson[1,4], Michael L. Mack[2,4],
Francis N. Bartlett III[3,4], and John M. Henderson[1,4]

[1] Michigan State University, Department of Psychology,
East Lansing MI 48824, USA
{dan, aaron, john}@eyelab.msu.edu
http://eyelab.msu.edu
[2] Michigan State University, Department of Computer Science,
East Lansing MI 48824, USA
mike@eyelab.msu.edu
[3] Michigan State University, Department of Zoology,
East Lansing MI 48824, USA
bartle47@msu.edu
[4] Michigan State University, Cognitive Science Program,
East Lansing MI 48824, USA

Abstract. An understanding of gaze control requires knowledge of the basic properties of eye movements during scene viewing. Because most of what we know about eye movement behavior is based on the viewing of images on computer screens, it is important to determine whether viewing in this setting generalizes to the viewing of real-world environments. Our objectives were to characterize eye movement behavior in the real world using head-mounted eyetracking technology and to illustrate the need for and development of automated analytic methods. Eye movements were monitored while participants searched for and counted coffee cups positioned within a cluttered office scene. Saccades were longer than typically observed using static displays, but fixation durations appear to generalize across viewing situations. Participants also made longer saccades to cups when a pictorial example of the target was provided in advance, suggesting a modulation of the perceptual span in accordance with the amount of information provided.

1   Introduction

The most highly resolved portion of one's visual field is derived by the foveal region of the retina where the cones are most densely packed. Because this region of high acuity is very small, corresponding only to about two degrees of visual angle, the eyes must be directed toward points of interest in a scene in order to encode visual details. Gaze shifts tend to occur at a rate of around three to four times per second with visual information extracted from the environment primarily when the direction of one's gaze is relatively stable. These periods of stability, called fixations, are separated by rapid movements of the eye, called saccades. A fundamental concern for those studying visual perception is the

L. Paletta et al. (Eds.): WAPCV 2004, LNCS 3368, pp. 83–99, 2005.

control of eye movements during scene viewing: What are the processes that determine the time spent engaged in fixations and govern the selection of targets for upcoming saccades?

Eye movement control is of interest to cognitive psychologists because eye movements are overt manifestations of the dynamic allocation of attention. While attention can be directed covertly away from the center of fixation, the natural tendency is for eye movements and attention to remain coupled. The most prevalent view of the relation between attention and eye movements is one where eye movements are preceded by shifts of attention to the location toward which the eyes are moving [1],[2],[3],[4],[5]. Sequential attention models such as the one proposed by Henderson [1] posit a relation between attention and eye movements that begins with attention allocated to the foveated stimulus. When processing at the center of fixation is complete or nearly complete, attention is disengaged and reallocated to a more peripheral location. This reallocation of attention coincides with the programming of an eye movement that brings the center of vision to the newly attended region of the visual field. In this view, questions of eye movement control are really questions of attentional selection because attentional selection and the targeting of saccadic eye movements are functionally equivalent.

Eye movement control is additionally of interest because of the degree of intelligence reflected in eye movement behavior. Early studies of picture viewing, for example, revealed a tendency for fixations to cluster in regions that were considered interesting and informative [6],[7],[8], (see [10] and [11] for reviews). Additionally, empty regions frequently did not receive fixation, suggesting that uninformative regions could be rejected as saccade targets on the basis of peripheral information. Subsequent research has been aimed at determining the roles of both visual and semantic informativeness in gaze control. A number of recent studies demonstrate a correlation between fixation densities and local image properties that can be represented in a wide range of statistics. For example, fixations tend to fall in regions that are high in local contrast and edge density compared to regions that do not receive fixation [12],[13],[14]. While subsequent research suggests that fixations are not drawn by local semantic information early in scene viewing [15],[16], (but see [17]), global properties of a scene, such as its meaning and overall spatial layout, can exert an influence as early as the first fixation. For example, participants in a search task were able to find target objects in scenes more quickly when a brief preview of the scene was presented in advance [18]. Finally, viewing patterns are influenced by the cognitive goal of the observer. In a classic study, Yarbus [9] monitored eye movements during the presentation of a picture of Repin's *An Unexpected Visitor*. The observed gaze patterns were modulated by the question posed to the viewer in advance, suggesting that the informativeness of scene regions is task-dependent.

While there have been numerous advances in the domain of eye movement control in scenes, most of what we know is derived from relatively artificial viewing situations. Eye movements are typically recorded while research participants view images presented on a computer monitor. The most commonly used eye tracking systems are considered stationary because the viewer's head position is maintained using an apparatus that includes a chin and forehead rest. Recent innovations in eyetracking technology, however, allow for the monitoring of eye movements as participants view real-world environments [19]. With these head-mounted eyetracking systems, eye movements are recorded along with a video image of the person's field of view provided by a camera positioned just above the eyes. Head-mounted eyetracking systems are advantageous because they allow researchers to study eye movements in more ecologically valid settings. For example, the field of view provided by a computer monitor is rather limited, as the display only extends to around 20° of visual angle. Head-mounted eyetrackers, on the other hand, allow eye movements to be monitored with a field of view that is unconstrained. In addition, allowing the head to move freely provides a more naturalistic viewing situation and allows for the study of eye movements during tasks that require interactions with the environment.

In the present study we wished to begin to address the question: How well does viewing on computer displays generalize to the viewing of natural environments? To investigate this question, we set up a simple search task in a real-world environment. Participants were asked to count coffee cups that were placed in various locations in a professor's office. The task provided the opportunity to examine a number of basic properties of eye movement behavior in a naturalistic setting.

First, we wished to determine the saccade lengths to objects of interest. Studies using static scenes on computer displays have predominately found mean saccade amplitudes in the $3-4°$ range [15],[16],[10]. Loftus and Mackworth [17], however, observed saccade lengths to objects that were as high as 7 degrees. Henderson and Hollingworth [10] suggested that this anomalous finding could reflect differences in stimuli. The Loftus and Mackworth [17] stimuli were line drawings of scenes that were relatively sparse compared to those used in other studies. The reduction in contours with stimuli of this kind would be expected to decrease the amount of lateral masking thereby increasing the distance from the center of vision from which useful information can be acquired (i.e., the perceptual span). While scene complexity was not manipulated in the present study, one would expect to find small saccades to cups because the room scene that was employed was cluttered with objects. Alternatively, saccade lengths to target objects might not generalize across viewing situations. In this case, saccade lengths could be longer in the natural environment than observed in previous studies using computer displays.

Second, because a complete understanding of eye movement control in real-world scenes will require a general knowledge of the basic properties of how the eyes move through a scene, we wished to generate fixation duration and saccade length distributions for the entire viewing episode. Mean fixation durations in search tasks have been reported as low as 275 ms [20] and 247 ms [16]; modal fixation durations around 220 ms have been reported in a search task using line drawings of scenes [16]. Mean and modal saccade lengths in picture viewing tasks have been reported as low as 0.5° and 2.4° respectively [10]. Recent data, however, suggest that smaller fixation durations and longer saccades may be more common in dynamic and naturalistic viewing situations [21]. Indeed, because the range of possible saccades in static displays is severely limited, determining saccade lengths in natural viewing situations is central to the question of the generalization of eye movement behavior.

Third, in addition to exploring the issue of generalizing extant eye movement data to the viewing of natural environments, we wished to begin exploring the concept of the perceptual span in real-world scenes. While the perceptual span has been extensively studied in the context of reading [20], very little has been done using scenes as stimuli; and to our knowledge there is nothing in the literature that addresses the perceptual span during the viewing of natural environments. In the present study, we wished to determine whether the perceptual span could be influenced by the precision of the representation of the target object. To investigate these issues, we compared the saccade lengths to two different sets of target items. One group of participants searched for coffee cups that came from a matched set and a second group searched for a mixture of coffee cups that varied in color, shape, and design. Importantly, the group that searched for matched cups were shown a picture of the target cup before the search began. If having a more precise representation of the target increases perceptual span, saccades to cups should be longer in the matched cup condition. This would be expected because participants in the matched cup condition would have more information available with which to reject regions as uninformative and select regions that are likely to have a cup.

Finally, while the present research was designed primarily to investigate a number of basic properties of eye movement control during the viewing of natural environments, it also represents the development of analytic tools for head-mounted eyetracking data. Despite the gains associated with using head-mounted eyetrackers, the utility of this technology is limited because data analyses are tedious and time-consuming. The greatest challenge is driven by the fact that the data are recorded in a reference frame that is dissociated from the world that the participant is viewing. With stationary eyetrackers, fixations are given in image plane coordinates that correspond to where the participant is looking on the computer display. The mapping from eyetracker space to display space is straightforward because the image plane and real-world coordinate systems coincide. Analyses can be automated because the locations of the fixations in the image plane coordinate system always correspond to the same locations on the display. Fixations are also given in image plane coordinates with head-mounted eyetrackers, but these coordinates correspond to varying locations in the real world because the image reflects a changing point of view. As a result, data analyses are often based on the hand scoring of video images that include a moving fixation cross indicating where the viewer is looking. The locations of fixations are manually determined by stepping one-by-one through the frames of video. This approach is extremely labor-intensive. Given that there are thirty frames for every second of viewing with a 30 Hz system (and more with faster systems), and every saccade and fixation must be determined, generating overall saccade length and fixation duration distributions with hand-scoring methods is usually impractical if not impossible. In this paper we report a method of automatically extracting the eyetracking data and we contrast the results from this method with the results derived from the hand-scoring method.

2   Present Research

2.1   Methods

Participants - Twenty-six Michigan State University undergraduates participated in exchange for course credit. All had normal or corrected-to-normal vision.

Stimuli - The room scene was a professor's office, approximately 8 x 12 meters, containing typical office items such as a desk, computer, chairs, file cabinets, and various shelving filled with books and paperwork (see Figure 1). The cups were placed in six locations throughout the room, spaced at relatively uniform distances from one another. The six cups used in the matched condition were all exactly the same. The other six cups were a variety of colors, shapes, and sizes.

Apparatus - Participants wore an ISCAN model ETL-500 head mounted eye-tracker. This eyetracker consists of a pair of cameras securely fastened to a visor that is strapped on the participant's head. One camera records the scene that the participant is currently viewing. The other, a camera sensitive to infrared light, records the movement of the participant's left eye. An infrared emitter that is housed within this camera illuminates the eye. Because this emitter is secured to the participant's head, the corneal reflection stays relatively stable across head and eye movements as the pupil moves. By measuring the offset between the pupil and the corneal reflection, it is possible to identify the location in the scene images that the participant is fixating. Since the visor is enclosed, participants were able to view 103° of visual angle horizontally, and 64° vertically. A plastic shield on the visor that is used to block infrared radiation when tracking the eyes outside was removed for this study, as it affects color vision.

In the ISCAN system, the scene and eye video are merged by a video multiplexer that takes input from both cameras simultaneously at 30 Hz. The multiplexed video, a composite video with the eye image on the left and scene image on the right, is then recorded by a Mini DV recorder. The final output is a scene video with a small cross on the image that corresponds to the center of fixation. The location of this cross is accurate to approximately one degree of visual angle under optimal conditions.

Procedure - Participants were tested individually and all testing was accomplished in one day to maintain control of the search environment. The session began in a laboratory adjacent to the professor's office. Upon arrival, participants were briefed about the experiment and equipment. The participants in the matched condition were additionally shown a pictorial example of the specific type of coffee cup they were to search for and count. Participants were then fitted with the eyetracker and calibrated. The calibration involved sequentially fixating five points arranged on a blackboard approximately 6 meters away. After calibration, they were walked to the professor's office. Before entering the office, the task instructions were repeated. Participants were then told to close their eyes and were led into the office by an experimenter. Upon entering the office, an experimenter placed the participants' feet over two pre-designated marks on the floor. This was done to ensure that each participant viewed the room from the same position. Once the participants were properly positioned, the office light was switched on and they were told to open their eyes and begin counting coffee cups out loud. They were given 15 seconds to count as many cups as they could find. Participants were aware of the limited amount of search time and were thus encouraged to find the cups as quickly as possible. When the time expired, an experimenter shut off the lights and told the participant to close their eyes. The participant was then led back to the laboratory for equipment removal and a memory test was administered to address questions that were not central to the topic of this paper. Once the memory test had been completed, participants were debriefed and allowed to leave.

2.2   Analysis

The objectives of the present study required two analytic methods. The first of these consisted of a hand-scoring method to determine saccade distances to each cup as well as the duration of the first fixation on each cup. The second consisted of developing and implementing an automated scoring method to eliminate subjectivity in separating fixations from saccades as well as for measuring the distance traveled during saccades and the durations of individual fixations. The hand-scored data were used as a benchmark to assist in developing the automated method of data analysis as well as in conjunction with the automatic analyses in order to converge on a common solution.

Before any scoring could begin, a number of intermediary steps were necessary to convert the video data into a workable format. The ISCAN ETL-500 data, in its raw form, is recorded onto 8mm digital videotape. At this point, the data are multiplexed and recorded onto a single tape to synchronize the separate video streams captured from the head-mounted eyetracker (HMET) eye camera and the scene video captured from the world camera.

The data are de-multiplexed into two video streams and recorded onto separate Mini DV (digital video) tapes via two Mini DV recorders. During this process, the ISCAN ETL-500 computer hardware and software interface utilize the pupil and corneal reflection from the eye video as well as the calibration sequence gathered during data collection to calibrate and display the point of regard of a given participant's eye onto the scene video (see Figure 1). This scene video is recorded onto DV tape along with the audio signal from the multiplexed Hi-8 video using one of the Mini DV recorders. The second recorder is used to record the video of the participant's eye.

Once the video stream has been separated and recorded onto DV tape, it is then transferred to computer via Adobe Premiere. The videos are then saved as

Fig. 1. An example of the scene video as it is output from the HMET hardware with a calibrated fixation cross indicating POR

avi files with a screen resolution of 720 x 480 pixels in their native DV format. This step maintains the original resolution of the DV videos as well as any compression associated with the DV format.

Scoring by Hand - Hand-scoring the video data involves locating the critical fixations within a series of digitized still images that were taken of the room scene while stepping frame-by-frame through the scene video output. The pixel distance from the launch fixation to the landing fixation was recorded for each cup that a participant counted. The pixel distances for the saccades to the cups were then expressed in terms of degrees of visual angle using a conversion factor determined for the still images. The conversion factor was derived by measuring the visual angle subtended by landmark features in the room (e.g., the length of shelves) from the participants' point of view. The visual angles subtended by the landmark features were then compared to the corresponding pixel distance in the still images to generate the conversion factor. Frame numbers for the onsets and offsets of the launching and landing fixations were also recorded to generate fixation durations for the fixations on the cups. Each video was scored by two experimenters and discrepancies were evaluated on a case-by-case basis.

Automated Scoring - Automated analysis of the eye and scene videos begins with syncing the videos to align corresponding frames. This is accomplished with an observer manually editing the videos so that the movements of the cross representing the point of regard (POR) in the scene video occur at the same frame number as the corresponding eye movements on the eye video.

Next, the center of mass of the pupil is recorded for each frame of the eye video. First, the pixels of the pupil region are extracted by converting the frames into binary images with a luminance threshold.

In many cases, the corneal reflection of the eye appears on the pupil. When this occurs, a poor extraction of the pupil region results from the threshold step explained above. With the area of the pupil not fully extracted, the pupil's center of mass cannot be correctly determined. This problem is overcome by fitting an ellipse to the pupil region and analyzing the resulting region created by the fit. The following conic ellipse model is fit to the boundary pixels of the pupil region with a least squares criterion:

$$ax^2 + bxy + cy^2 + dx + ey + f = 0 \qquad (1)$$

From this fit, the center of mass, area, and vertical diameter of the pupil are calculated and recorded (see Figure 2).

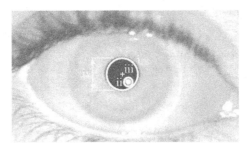

Fig. 2. Sample frame from eye video. i. Pupil ii. Corneal reflection iii. Center of mass of pupil iv. Vertical diameter of pupil

At this point, we determine when fixations, saccades, and blinks occur. A threshold is set for change in pupil area and change in pupil diameter to determine when a blink occurs. Fortunately, the change in pupil area and pupil diameter during a blink is much faster than during pupil dilation and constriction, so a threshold is easily set. Next, saccades are determined with a two-step process. Using the acceleration of the pupil's center of mass, a liberal decision of when a saccade occurs can be made with another threshold. By first using acceleration to find saccades, slower movements of the eye that keep the point of regard stable on the world while the head is in motion can be recognized and scored as fixations. The next step involves using a two-stage threshold for the velocity of the pupil's center of mass (see Figure 3). The first threshold finds the peak velocity of a saccade. Once this threshold has been exceeded, the frames before and after the current frame are progressively examined to find the tails of the saccade. This analysis results in each frame of the eye video labeled as part of a fixation, saccade, or blink.

Fixations are extracted from the frame labels by grouping together contiguous frames labeled as part of a fixation. Fixation durations are calculated by dividing the number of frames in a fixation by the frame rate of the eye video (e.g., a fixation with 5 frames recorded at 30 frames per second would yield a fixation duration of 167 ms). Saccade lengths are calculated by determining the Euclidian distance traveled by the center of mass in the eye video from one fixation to the next. This pixel distance is then converted to degrees of visual angle using

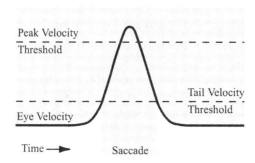

Fig. 3. A common velocity pattern during a saccade is shown by the black line. First, the middle of a saccade is found with a peak velocity threshold (top dotted line). Once this threshold has been exceeded, neighboring frames are progressively included in the saccade until the frames' velocities drop below the smaller tail velocity threshold (bottom dotted line). This algorithm allows for more accurate determination of the entire saccade

a conversion factor determined by measuring the distance traveled by the eye during calibration.

To locate the fixations corresponding to cups, the scene videos (as output from the HMET hardware) were divided evenly among three independent observers. Cup fixations were located utilizing the participants' audible responses to identify where the cup was counted, then stepping backward frame-by-frame from this point until a corresponding cup fixation was located in the video as indicated by the POR fixation cross. The frame number at which this fixation occurred was recorded into a spreadsheet. This allowed the frame locations at which cup fixations occurred to be noted in the fixation-listing file generated by the automated scoring method. The saccade lengths to the cups were the saccade lengths determined for the fixation on the cups in the fixation-listing file.

3   Results

Six (out of 26) participants were excluded from the analyses. One of them was removed due to an error during data collection resulting in data that was not analyzable. An additional participant was removed because it appeared that this person misunderstood the instructions. The other three participants were eliminated due to increased noise in the tracking caused by extraneous reflections in the eye image - making the data very difficult to analyze by hand, and impossible to cross-evaluate via the automated data collection software. All of the following analyses were based on the remaining 20 participants, 9 in the matched cup condition and 11 in the mixed cup condition.

3.1   Saccade Lengths and Fixation Durations for Cups

The mean saccade lengths to the cups were 10.7° and 11.9° for the automatic and hand-scored methods respectively. Table 1 shows the mean saccade lengths to the cups by search condition for each of the scoring methods. The saccade lengths to the cups in the two search conditions were compared using a one-way analysis of variance (ANOVA). The mean saccade length to the cups was reliably greater in the matched cup condition than the mixed cup condition using the hand-scoring method, $F(1, 18) = 5.36$, $MSE = 25.45$, $p < .05$, and marginally greater using the automated method, $F(1, 18) = 3.85$, $MSE = 16.23$, $p < .10$. The mean fixation durations on the cups were 267 ms and 284 ms for the automated and hand-scored methods respectively. Table 1 also shows the mean fixation durations on cups by search condition. Although fixation durations were numerically greater in the mixed condition with both methods, these differences were not reliable ($Fs < 1$).

Table 1. Mean saccade lengths and fixation durations for cups derived by the automated and hand-scoring methods

	Condition	Automated	Hand-scored
Saccade length (deg)	Matched	8.8	9.0
	Mixed	12.3	14.2
Fixation Durations (msec)	Matched	258	272
	Mixed	275	294

3.2   Saccade Length Distribution

Figure 4 shows the overall distribution of saccade lengths generated by the automated scoring method for the entire viewing episode. The mean and mode of the distribution were 12.8 and 6.8 respectively. The mean saccade lengths did not differ between the two search conditions, $F(1, 1087) = 2.37$, $MSE = 107.77$, $p = .12$.

The automated and hand-scoring methods were cross-evaluated by comparing the set of saccade lengths that were measured using both methods. That is, the saccade lengths to the cups derived by the automated method were compared to the same saccades derived by the hand-scored method. Figure 5 shows the distributions of saccade lengths to cups derived from each of the two methods. As can be seen in the figure, the mode was higher using the automatic method (7.4°) than it was using the hand-scoring method (3.3°). The two distributions were compared using a repeated measures ANOVA. The overall mean based on hand-scoring (11.7°) was greater than the overall mean derived automatically (10.6°), $F(1, 88) = 6.38$, $MSE = 8.35$, $p < .05$. However, the difference between the means was quite small as was the size of the effect (partial eta squared was .068).

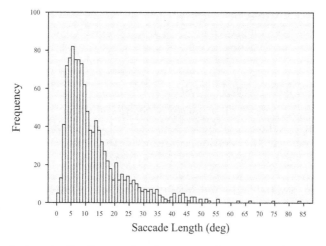

Fig. 4. Saccade length distribution for the entire viewing episode derived from the automatic scoring method

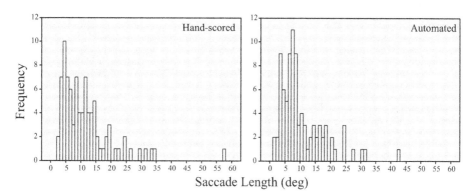

Fig. 5. Distributions of saccade lengths to the cups for the automatic and hand-scoring methods

Overall, the agreement between scoring methods was quite good. To begin, the saccade lengths derived from the two methods were reliably correlated ($r = .886$, $p < .001$). To determine quantitatively the nature of the relation between the two measures, the automatic scores were entered into a linear regression analysis with hand scores used as the predictor variable. The slope and intercept terms were both reliable, $b_1 = 0.76$, $p < .001$, and $b_0 = 1.75$, $p < .01$, respectively. A visual inspection of the distributions suggested that the largest difference between the distributions was the elevated number of saccades in the 7° range using the automated method. There were just as many short saccades in the automated distribution as there were in the hand-scored distribution. Thus, the mean was higher in the hand-scored distribution because of the increased number of longer saccades as opposed to a decreased number of shorter saccades. Indeed, the regression suggests that the difference between the methods increases as saccades lengths increase with the automated method producing smaller estimates of saccades in the upper range of the distribution.

3.3   Fixation Duration Distribution

Figure 6 shows the overall distribution of fixation durations derived by the automated scoring method. The mean and mode of the distribution were 210 ms and 133 ms respectively. Fixation durations were greater overall in the mixed cup condition (mean = 220 ms) than in the match cup condition (mean = 199 ms), $F(1, 1109) = 10.31$, $MSE = 12334.59$, $p < .01$. As in the saccade length analysis, the automated and hand-scoring methods were cross-evaluated by comparing the fixation durations on the cups derived by the automated method to the same fixations derived by the hand-scored method. Figure 7 shows the distributions of fixation durations on cups derived by each of the two methods. As can be seen in the figure, the mode was 200 ms using both scoring methods. The two distributions were compared using a repeated measures ANOVA. The overall mean based on hand-scoring (281 ms) was marginally greater than the overall mean derived automatically (265 ms), $F(1, 88) = 2.70$, $MSE = 4053$, $p = .10$.

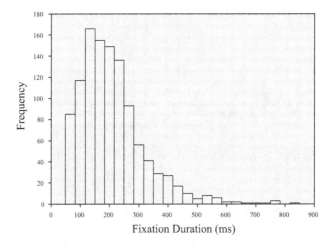

Fig. 6. Fixation duration distribution for the entire viewing episode derived from the automatic scoring method

Overall, the agreement between scoring methods was quite good. The fixation durations derived by the two methods were reliably correlated ($r = .748$, $p < .001$). Again, to determine quantitatively the nature of the relationship between the two measures, the automatic scores were entered into a linear regression analysis with hand scores used as the predictor variable. The intercept term was not reliable when it was included in the model, $b_0 = 24.67$, $p = n.s.$, and the slope was 0.93 ($p < .001$) when the model was run without an intercept term. Visually, the largest difference between the distributions was the higher peak with the hand-scored method and the fact that there were twice as many fixations below 200 ms using the automated method. There were, however, no major systematic differences between the distributions.

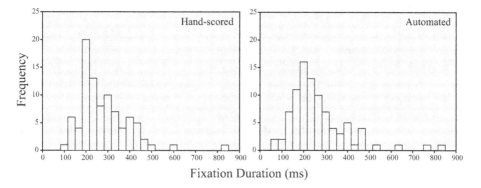

Fig. 7. Distributions of fixation durations on cups for the automatic and hand-scoring methods

4   Discussion

The main objective of this chapter was to begin to explore how eye movements are used to understand real-world environments. Up until now, the majority of research has been directed toward eye movements over scenes depicted in static displays. We believe that an important next step in scene perception research involves observing and understanding how the visual system behaves in real-world settings. To begin to examine this issue, we employed a visual search task, a task that is commonly used with static displays. Within the context of this task, we examined basic eye movement parameters to evaluate the generality of these measures in naturalistic settings.

The saccade length distributions generated by the present study provide a useful tool for evaluating real-world eye movement behavior. Our results yielded a greater number of long saccades than one might predict from earlier work. Both the saccade lengths to cups, as well as the distribution of saccade lengths overall were considerably larger than those reported in studies using static displays. The critical question is whether the differences in saccade length distributions are due to real differences in eye movement behavior versus due to differences in the equipment used to measure the behavior. A criticism of head-mounted eyetracking is that the spatial resolution does not allow for the detection of very small intra-object saccades. With the most precise stationary eyetrackers, however, saccades as small as 0.5° are very frequent. Therefore, the abundance of longer saccades in our data could be driven by an insensitivity to smaller saccades.

There are a number of reasons to believe that this is not the case. First, the automated method allowed us to capture the eye movement behavior directly from the eye video rather than at a post processing stage. Since error would be likely to occur at each stage of processing, the spatial resolution issues that are present in the scene video are likely to be amplified relative to that provided by the eye video alone. In addition, the difference between automated and hand-scoring methods was greater at the high end of the distribution. The agreement between methods at the low end of the distribution suggests that even the scene video captures the lower end of the distribution reasonably well.

Second, if the mode of the distribution was elevated by an insensitivity to small saccades, we would expect the saccade length distribution to be truncated at the location of the lower limit of our equipment's spatial resolution. However, the pre-modal frequency of saccade lengths is greater than one would expect if small saccades are being lumped into fixations, and the post-modal tail of the distribution extends much further than it does in distributions derived from stationary eyetrackers on static displays. Visually, the similarity of the shape of this distribution to that found using static displays suggests that the distribution has actually shifted - reflecting an overall difference in viewing strategy between real-world settings and static displays. In addition to saccade lengths, we compared distributions of fixation durations. Fixation durations in the present study were appreciably smaller than those reported using stationary eyetrackers and static displays. However, as with the saccade length distribution, it is important to determine the extent to which differences might be attributed to equipment rather than behavior. One problem that arises in the generation of fixation distributions results from the abundance of head movements that occur during the viewing episode. Though the algorithm used by the automated scoring method was designed to robustly dismiss the slow eye movements that stabilize the POR in real-world coordinates while the head is moving, the potential for inappropriately splitting fixations remains. The fact that there were twice as many short saccades with the automatic method compared to the hand-scoring method supports this conclusion. This would not be an issue for the hand-scoring method because this is based on the stability of the POR in world-centered coordinates as reflected in the scene video. The automated scoring method, on the other hand, is based entirely on the stability or instability of the eye independent of this frame of reference.

A second issue for interpreting fixation durations comes from the limited temporal resolution of the HMET system. While the most resolute stationary eyetrackers are capable of sampling at rates greater than 1000 Hz, the video output of the HMET system is captured at a rate of 30 Hz (which corresponds to 33 ms frame samples). This difference would serve to artificially shorten our fixation durations because fixations that terminate just outside the sample window would be trimmed by as much as 33 ms. Given this amount of measurement error, we can only say with confidence that the mode lies between 133 ms and 166 ms and the mean lies between 210 ms and 243 ms. While this mean is comparable to means found using stationary eyetrackers and static displays, the frequency of smaller fixations could reflect real differences in viewing behavior between our task and the tasks used in earlier studies. This could reflect the influence of time pressure, but further research is necessary to disentangle this result.

In addition to issues of the generality of eye-movement behavior during real-world viewing, our experiment was designed to determine whether differences in the specificity of the stimulus of interest modulates the perceptual span. The hypothesis was that by providing an example of the cup in advance the visual system would be better tuned to detect features in the environment that are likely to belong to the search target, thereby increasing the perceptual span. Under the assumption that saccade lengths reflect perceptual span, the saccades to the cups were expected to be greater in the matched cup condition. The saccade lengths to the cups observed with the automated and hand-scoring methods were both in agreement with this prediction. While this result lends support to the idea that the visual system is designed to adapt to specific task-related constraints, the fact that cups were the same color in the matched condition and of varied colors in the mixed condition prohibits a strong interpretation because the items were not matched for saliency. The fact that overall fixation durations were greater in the mixed condition is consistent with this possibility. Nevertheless, the reliable difference in saccade lengths between conditions demonstrates an ability to detect differences in perceptual span with the present methodology and suggests a promising avenue for future research.

5   Summary

The control of eye movements during scene perception is a topic that continues to gain interest in the scientific community. Our goal in this chapter was not only to begin to test the generality of eye movements in real-world environments but also to illustrate the advantages and challenges associated with head-mounted eyetracking. In the study of scene perception, what we really want to know is how the eyes behave during naturalistic viewing situations, where head movements are allowed and the field of view is unconstrained. Yet in practice we often use relatively constrained settings to make inferences about how people view the real world. One of the unique abilities that a head-mounted eyetracker affords over a stationary eyetracker is complete freedom of movement on the part of the participant. One could therefore argue that this methodology results in a more accurate portrayal of eye movement behavior in the real world. However, the complexity of the data analysis poses a considerable challenge. Eye movements are complex enough when the head is stationary and the view is constrained. Discerning meaningful patterns from this complex behavior is compounded by the dissociation of reference frames introduced by the freedom of head movement. As a result, the utility of HMET depends greatly on the advancement of analytical methods to deal with the added complexity.

The differences observed in the present study relative to those found using static displays suggests the enterprise of advancing these methods is warranted, and the degree of convergence between the automated and hand-scoring methods suggests the automated methodology is a promising start. There are a number of avenues that remain to be explored. For example, quantifying head movements has the potential to increase the accuracy with which saccades and fixations are determined because head movements seemed to occur within every observed video frame in the present study. However, because the abundance of head movement could be a product of the time pressure imposed in the current task, future methods should be tested in the context of alternative tasks. While utilizing HMET in a search task provides an initial glance, exploration with other tasks is necessary to provide a complete picture of eye movement control in real-world settings.

In conclusion, the data reported here suggest both similarities and differences between the viewing of natural environments and the viewing of static displays. Saccades lengths were longer than typically observed using static displays, but fixation durations appear to generalize across viewing situations. Finally, the scaling of saccades with the precision of the search target's representation provides yet another example of the visual system's ability to adapt and make efficient use of the information it is given.

Acknowledgments

This work was supported by the National Science Foundation (BCS-0094433) and the Army Research Office (W911NF-04-1-0078), and by a National Science Foundation IGERT graduate training grant (ECS-9874541). The opinions expressed in the article are those of the authors and do not necessarily represent the views of the department of the Army or other governmental organizations. Reference to or citations of trade or corporate names do not constitute explicit or implied endorsement of those entities or their products by the authors or the Department of the Army. Facilities for this research were supported by the Center for the Integrated Study of Vision and Language. We wish to thank Karl Bailey, Monica Castelhano, Dirk Colbry, and Nan Zhang for their contributions to the work presented.

References

1. Henderson, J. M.: Visual attention and eye movement control during reading and picture viewing. In K. Rayner (Ed.), Eye movements and visual cognition: Scene perception and reading (pp. 260-283). New York Springer-Verlag (1992)
2. Henderson, J. M., Pollatsek, A., & Rayner, K.: Covert visual attention and extrafoveal information use during object identification. Perception & Psychophysics, 45, (1989) 196-208
3. Hoffman, J. E., & Subramaniam, B. (1995). The role of visual attention in saccadic eye movements. Perception & Psychophysics, 57, 787-795.

4. Kowler, E., Anderson, E., Dosher, B., & Blaser, E. (1995). The role of attention in the programming of saccades. Vision Research, 35, 1897-1916.
5. Shepherd, M., Findlay, J. M., & Hockey, R. J. (1986). The relationship between eye movements and spatial attention. The Quarterly Journal of Experimental Psychology, 38A, 475-491.
6. Antes, J.R. (1974). The time course of picture viewing. Journal of Experimental Psychology, 103, 62-70.
7. Buswell, G. T. (1935). How people look at pictures. Chicago: University of Chicago Press.
8. Mackworth, N. H., & Morandi, A. J. (1967). The gaze selects informative details within pictures. Perception and Psychophysics, 2, 547-552.
9. Yarbus, A.L. (1967). Eye Movements and Vision. New York: Plenum Press.
10. Henderson, J. M., & Hollingworth, A. (1998). Eye movements during scene viewing: An overview. Eye Guidance in Reading and Scene Perception. Edited by: Underwood, G., Elsevier Science Ltd. 269-293.
11. Henderson, J. M. (2003). Human gaze control in real-world scene perception. Trends in Cognitive Sciences, 7, 498-504.
12. Mannan, S. K., Ruddock, K. H., & Wooding, D. S. (1996). The relationship between the locations of spatial features and those of fixations made during visual examination of briefly presented images. Spatial Vision, 10, 165-188.
13. Mannan, S. K., Ruddock, K. H., & Wooding, D. S. (1997). Fixation patterns made during brief examination of two-dimensional images. Perception, 26, 1059-1072.
14. Parkhurst, D., & Niebur, E. (2003). Scene content selected by active vision. Spatial Vision, 16, 125-154.
15. De Græf, P., Christiæns, D., & d'Ydewalle, G. (1990). Perceptual effects of scene context on object identification. Psychological Research, 52, 317-329.
16. Henderson, J. M., Weeks, P. A., Jr., & Hollingworth, A. (1999). The effects of semantic consistency on eye movements during complex scene viewing. Journal of Experimental Psychology: Human Perception and Performance, 25, 210-228.
17. Loftus, G. R., & Mackworth, N. H. (1978). Cognitive determinants of fixation location during picture viewing. Journal of Experimental Psychology: Human Perception and Performance, 4, 565-572.
18. Castelhano, M. S., & Henderson, J. M. (2003). Flashing scenes and moving windows: An effect of initial scene gist on eye movements [Abstract]. Journal of Vision, 3(9), 67a, http://journalofvision.org/3/9/67/, doi:10.1167/3.9.67.
19. Land, M. F., & Hayhoe, M. (2001). In what ways do eye movements contribute to everyday activities? Vision Research, 41, 3559-3565.
20. Rayner, K. (1998). Eye movements in reading and information processing: 20 years of research. Psychological Bulletin, 124, 372-422.
21. Pelz, J. B., Canosa, R., Lipps, M., Babcock, J., & Rao, P. (2003). Saccadic targeting in the real world [Abstract]. Journal of Vision, 3(9), 310a, http://journalofvision.org/3/9/310/, doi:10.1167/3.9.310.

The Computational Neuroscience of Visual Cognition: Attention, Memory and Reward

Gustavo Deco

Institució Catalana de Recerca i Estudis Avançats (ICREA),
Universitat Pompeu Fabra, Dept. of Technology, Computational Neuroscience,
Passeig de Circumval.lació, 8 - 08003 Barcelona, Spain

Abstract. Cognitive behaviour requires complex context-dependent processing of information that emerges from the links between attentional perceptual processes, working memory and reward-based evaluation of the performed actions. We describe a computational neuroscience theoretical framework which shows how an attentional state held in a short term memory in the prefrontal cortex can by top-down processing influence ventral and dorsal stream cortical areas using biased competition to account for many aspects of visual attention. We also show how within the prefrontal cortex an attentional bias can influence the mapping of sensory inputs to motor outputs, and thus play an important role in decision making. This theoretical framework incorporates spiking and synaptic dynamics which enable single neuron responses, fMRI activations, psychophysical results, and the effects of damage to parts of the system, to be explicitly simulated and predicted. This computational neuroscience framework provides an approach for integrating different levels of investigation of brain function, and for understanding the relations between them.

Keywords: visual attention, computational neuroscience, biased competition, theoretical model.

1   Introduction

To understand how the brain works, including how it functions in vision it is necessary to combine different approaches, including neural computation. Neurophysiology at the single neuron level is needed because this is the level at which information is exchanged between the computing elements of the brain. Evidence from neuropsychology is needed to help understand what different parts of the system do and what each part is necessary for. Neuroimaging is useful to indicate where in the human brain different processes take place, and to show which functions can be dissociated from each other. Knowledge of the biophysical and synaptic properties of neurons is essential to understand how the computing elements of the brain work, and therefore what the building blocks of biologically realistic computational models should be. Knowledge of the anatomical and functional architecture of the cortex is needed to show what types of neuronal

L. Paletta et al. (Eds.): WAPCV 2004, LNCS 3368, pp. 100–117, 2005.

network actually perform the computation. And finally the approach of neural computation is needed, as this is required to link together all the empirical evidence to produce an understanding of how the system actually works. This review utilizes evidence from some of these disciplines to develop an understanding of how vision is implemented by processing in the brain, focusing on visual attentional mechanisms. A theoretical framework that fulfils these requirements can be obtained by developing explicit mathematical neurodynamical models of brain function based at the level of neuronal spiking and synaptic activity [24]. In this article we show how this approach is being used to produce a unified theory of attention and working memory, and how these processes are influenced by rewards to influence decision making.

2   Visual Attention: Experimental Studies

2.1   Single Cell Experiments

The dominant neurobiological hypothesis to account for attentional selection is that attention serves to enhance the responses of neurons representing stimuli at a single relevant location in the visual field. This enhancement model is related to the metaphor for focal attention in terms of a spotlight [27, 28]. This metaphor postulates a spotlight of attention which illuminates a portion of the field of view where stimuli are processed in higher detail while the information outside the spotlight is filtered out. According to this classical view, a relevant object in a cluttered scene is found by rapidly shifting the spotlight from one object in the scene to the next one, until the target is found. Therefore, according to this assumption the concept of attention is based on explicit serial mechanisms. There exists an alternative mechanism for selective attention, the *biased competition* model [12, 13, 14]. According to this model, the enhancement of attention on neuronal responses is understood in the context of competition among all of the stimuli in the visual field. The *biased competition* hypothesis states that the multiple stimuli in the visual field activate populations of neurons that engage in competitive mechanisms. Attending to a stimulus at a particular location or with a particular feature biases this competition in favor of neurons that respond to the location or the features of the attended stimulus. This attentional effect is produced by generating signals within areas outside the visual cortex which are then fed back to extrastriate areas, where they bias the competition such that when multiple stimuli appear in the visual field, the cells representing the attended stimulus "win", thereby suppressing cells representing distracting stimuli. According to this line of work, attention appears as an emergent property of competitive interactions that work in parallel across the visual field. Reynolds et al. [22] first examined the presence of competitive interactions in the absence of attentional effects, making the monkey attend to a location far outside the receptive field of the neuron they were recording. They compared the firing activity response of the neuron when a single reference stimulus was located within the receptive field to the response when a probe stimulus was added to the visual field. When the probe was added to the field, the activity of the neuron was shifted towards the activity level that would have been evoked had the probe appeared alone. When the reference is an effective stimulus (high response) and the probe is an ineffective stimulus (low response) the firing activity is suppressed after adding the probe. In contrast, the response of the cell increased when an effective probe stimulus was added to an ineffective reference stimulus. The study also tested attentional modulatory effects independently by repeating the same experiment with the difference that the monkey attended to the reference stimulus within the receptive field of the recorded neuron. The effect of the attention on the response of the V2 neuron was to almost compensate the suppressive or excitatory effect of the probe. That is, if the probe caused a suppression of the activity response to the reference when the attention was outside the receptive field, then attending to the reference restored the neuron's activity to the level corresponding to the case of the reference stimulus alone. Similarly, if the probe stimulus had increased the neuron's level of activity, attending to the reference stimulus compensates the response by shifting the activity to the level that had been recorded when the reference was presented alone.

2.2   fMRI Experiments

The experimental studies of Kastner et al. [17, 18] show that when multiple stimuli are present simultaneously in the visual field, their cortical representations within the object recognition pathway interact in a competitive, suppressive fashion. The authors also observed that directing attention to one of the stimuli counteracts the suppressive influence of nearby stimuli. These experimental results were obtained by applying the functional magnetic resonance imaging (fMRI) technique in humans. The authors designed an experiment and different conditions were examined. In the first experimental condition the authors tested the presence of suppressive interactions among stimuli presented simultaneously within the visual field in the absence of directed attention, in the second experimental condition they investigated the influence of spatially directed attention on the suppressive interactions, and in the third condition they analyzed the neural activity during directed attention but in the absence of visual stimulation. The authors observed that, because of the mutual suppression induced by competitively interacting stimuli, the fMRI signals were smaller during the simultaneous presentations than during the sequential presentations. In the second part of the experiment there were two main factors: presentation condition (sequential versus simultaneous) and directed attention condition (unattended versus attended). The average fMRI signals with attention increased more strongly for simultaneously presented stimuli than the corresponding signals for sequentially presented stimuli. Thus, the suppressive interactions were partially cancelled out by attention.

2.3   Psychophysical Experiments: Visual Search

We now concentrate on the macroscopic level of psychophysics. Evidence for different temporal behaviours of attention in visual processing come from psychophysical experiments using visual search tasks where subjects examine a display containing randomly positioned items in order to detect an *a priori* defined target. All other items in the display which are different from the target serve the role as distractors. The relevant variable tipically measured is search time as a function of the number of items in the display. Much work has been based on two kinds of search paradigm: feature search, and conjunction search. In a feature search task the target differs from the distractors in one single feature, e.g. only colour. In a conjunction search task the target is defined by a conjunction of features and each distractor shares at least one of those features with the target. Conjunction search experiments show that search time increases linearly with the number of items in the display, implying a serial process. On the other hand, search times in a feature search can be independent of the number of items in the display. Quinlan and Humphreys [21] analyzed feature search and three different kinds of conjunction search, namely: standard conjunction search and two kinds of triple conjunction with the target differing from all distractors in one or two features respectively. Let us define the different kinds of search tasks by using a pair of numbers m and n, where m is the number of distinguishing feature dimensions between target and distractors, and n is the number of features by which each distractor group differs from the target. Using this terminology, feature search corresponds to a 1,1-search; a standard conjunction search corresponds to a 2,1-search; a triple conjunction search can be a 3,1 or a 3,2-search depending of whether the target difffers from all distractor groups by one or two features respectively. Quinlan and Humphreys [21] showed that in feature search (1,1), the target is detected in parallel across the visual field. They also show that the reaction time in both standard conjunction search and triple conjunction search conditions is a linear function of the display size. The slope of the function for the triple conjunction search task can be steeper or relatively flat, depending upon whether the target differs from the distractors in one (3,1) or two features (3,2), respectively.

3   A Neurodynamical Model of Attention

The overall systemic representation of the model is shown in Fig.1. The system is essentially composed of six modules (V1, V2-V4, IT, PP, v46, d46), structured such that they resemble the two known main visual paths of the mammalian visual cortex: the *what* and *where* paths [9,16,10]. These six modules represent the minimum number of components to be taken into account within this complex system in order to describe the desired visual attention mechanism. Information from the retino-geniculo-striate pathway enters the visual cortex through areas V1-V2 in the occipital lobe and proceeds into two processing streams. The occipital-temporal stream (*what* pathway) leads ventrally through V4 and IT (inferotemporal cortex) and is mainly concerned with object recognition, independently of position and scaling. The occipito-parietal stream (*where* pathway) leads dorsally into PP (posterior parietal) and is concerned with the location of objects and the spatial relationships between objects. The model considers that feature attention biases intermodular competition between V4 and IT, whereas spatial attention biases intermodular competition between V1, V4 and PP.

The ventral stream consists of four modules: V1, V2-V4, IT, and a module v46 corresponding to the ventral area 46 of the prefrontal cortex, which maintains the short-term memory of the recognized object or generates the target object in a visual search task. The module V1 is concerned with the extraction of simple features (for example bars at different locations, orientations and size). It consists of pools of neurons with Gabor receptive fields tuned at different positions in the visual field, orientations and spatial frequency resolutions. The V1 module contains P x P hypercolumns that cover the N x N pixel scene. Each hypercolumn contains L orientation columns of complex cells with K octave levels corresponding to different spatial frequencies. This V1 module inputs spatial and feature information up to the dorsal and ventral streams. Also, there is one inhibitory pool interacting with the complex cells of all orientations at each scale. The inhibitory pool integrates information from all the excitatory pools within the module and feedbacks unspecific inhibition uniformly to each of the excitatory pools. It mediates normalizing lateral inhibition or competitive interactions among the excitatory cell pools within the module. The module IT is concerned with the recognition of objects and consists of pools of neurons which are sensitive to the presence of a specific object in the visual field. It contains C pools, as the network is trained to search for or recognize C particular objects. The V2-V4 module serves primarily to pool and channel the responses of V1 neurons to IT to achieve a limited degree of translation invariance. It also mediates a certain degree of localized competitive interaction between different targets. A lattice is used to represent the V2-V4 module. Each node in this lattice has L x K assemblies as in a hypercolumn in V1. Each cell assembly receives convergent inputs from the cell assemblies of the same tuning from an M x M hypercolumn neighborhood in V1. The feedforward connections from V1 to V2-V4 are modeled with convergent Gaussian weight function, with symmetric recurrent connection.

The dorsal stream consists of three modules: V1, PP and d46. The module PP consists of pools coding the position of the stimuli. It is responsible for mediating spatial attention modulation and for updating the spatial position of the attended object. A lattice of N x N nodes represents the topographical organization of the module PP. Each node on the lattice corresponds to the spatial position of each pixel in the input image. The module d46 corresponds to the dorsal area 46 of the prefrontal cortex that maintains the short term spatial memory or generates the attentional bias for spatial location. The prefrontal areas 46 (modules v46 and d46) are not explicitly modeled. Feedback connections between these areas provide the external top-down bias that specifies the task. The feedback connection from area v46 to the IT module specifies the target object in a visual search task. The feedback connection from area d46 to the PP module generates the bias to a targeted spatial location.

The system operates in two different modes: the learning mode and the recognition mode. During the learning mode, the synaptic connections between V4

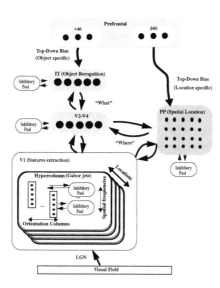

Fig. 1. Architecture of the neurodynamical approach. The system is essentially composed of six modules structured such that they resemble the two known main visual paths of the visual cortex

and IT are trained by means of Hebbian learning during several presentations of a specific object. During the recognition mode there are two possibilities of running the system. First, an object can be localised in a scene (visual search) by biasing the system with an external top-down component at the IT module which drives the competition in favour of the pool associated with the specific object to be searched. Then, the intermodular attentional modulation V1-V4-IT will enhance the activity of the pools in V4 and V1 associated with the features of the specific object to be searched. Finally, the intermodular attentional modulation V4-PP and V1-PP will drive the competition in favour of the pool localising the specific object. Second, an object can be identified (object recognition) at a specific spatial location by biasing the system with an external top-down component at the PP module. This drives the competition in favour of the pool associated with the specific location such that the intermodular attentional modulation V4-PP and V1-PP will favour the pools in V1 and V4 associated with the features of the object at that location. Intermodular attentional modulation V1-V4-IT will favour the pool that recognized the object at that location.

The activity of each computational unit is described by a dynamical equation derived from the mean field approximation (see Appendix and [24] for details).

3.1   Simulation of Single-Cell Experiments

Deco et al. [5, 4] presented simulations of the experiments by Reynolds *et al.* on single cell recording in V4 neurons in monkeys. They studied the dynamical behavior of the cortical architecture presented in the previous section by numerically solving the system of coupled differential equations in a computer simulation. They analyzed the firing activity of a single pool in the V4 module which was highly sensitive to a vertical bar presented in its receptive field (effective stimulus) and poorly sensitive to a 75 degrees oriented bar presented in its receptive field (ineffective stimulus). Following the experimental setup of the work by Reynolds *et al.*, they calculated the evolution of the firing activity of a V4 pool under four different conditions: 1) single reference stimulus within the receptive field; 2) single probe stimulus within the receptive field; 3) reference and probe stimulus within the receptive field without attention; 4) reference and probe stimuli within the receptive field and attention directed to the spatial location of the reference stimulus. In the simulations, the attention was directed to the reference location by setting a top-down attentional bias in PP to the location of the reference stimulus. In the unattended condition, the external top-down bias was set equal to zero everywhere. Comparing with the experimental results, the same qualitative behavior is observed for all experimental conditions analyzed. The competitive interactions in the absence of attention are due to the intramodular competitive dynamics at the level of V1 (i.e. the suppressive and excitatory effects of the probe). The modulatory biasing corrections in the attended condition are caused by the intermodular interactions between V1 and PP pools, and PP pools and prefrontal top-down modulation.

3.2   Simulations of fMRI Data

The dynamical evolution of activity at the cortical area level, as evidenced in the behaviour of fMRI signals in experiments with humans, can be simulated in the framework of the present model by integrating the pool activity in a given area over space and time. The integration over space yields an average activity of the considered brain area at a given time. With respect to the integration over time, it is performed in order to simulate the temporal resolution of fMRI experiments. We simulated fMRI signals from V4 under the experimental conditions defined by Kastner *et al.* [18]. In order to simulate the data by Kastner *et al.*, we also used four complex images similar to the ones these authors used in their work. These images were presented as input images in four nearby locations in the upper right quadrant. The neurodynamics is solved through an interactive process. Stimuli were shown in the two above mentioned conditions: sequential and simultaneous (see Fig. 2a). In the SEQ condition, stimuli were presented separately in one of the four locations for 250 ms. In the SIM condition, the four stimuli appeared simultaneously for 250 ms and with equal blank intervals between each other. The order of the stimuli and location was randomized. Two attentional conditions were simulated: an unattended condition, during which no external top-down bias from prefrontal areas was present (i.e. $I_{ij}^{PP,A}$ is zero everywhere) and an attended condition that was defined 10s before the onset of visual presentations (expectation period EXP) and continued during the subsequent 10s block.

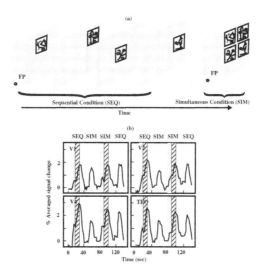

Fig. 2. (a) Experimental design of Kastner et al (1999). (b) Computer simulations of fMRI signals in visual cortex. Grey shade areas indicate the expectation period, striped areas the attended presentations and blocks without shading correspond to unattended condition

The attended condition was implemented by setting $I_{ij}^{PP,A}$ equal to 0.07 for the locations associated with the lowest left stimulus and zero elsewhere. Fig.2b shows the results of the computational simulations [4] for a sequential simulation block: BLK-EXP-SEQ(attended)-BLK-SIM-BLK-EXP-SIM(attended). As in the experiments of Kastner *et al.* [18], these simulations show that the fMRI signals were smaller in magnitude during the SIM than during the SEQ presentations in the unattended conditions because of the mutual suppression induced by competitively interacting stimuli. On the other hand, the average fMRI signals with attention increased more strongly for simultaneously presented stimuli than the corresponding ones for sequentially presented stimuli. Thus, the suppressive interactions were partially cancelled out by attention. Finally, during the expectation period activity increased in the absence of visual presentations and further increased after the onset of visual stimuli. The theoretical data describe quite well the qualitative behaviour of the experiments.

3.3   Simulation of Visual Search Tasks

Deco and Zihl [10] and Rolls and Deco [24] extended the neurodynamical model to account for the different slopes observed experimentally in complex conjunction visual search tasks. Figure 3 shows the overall architecture of the extended model. The input retina is given as a matrix of visual items. The location of each item on the retina is specified by two indices (ij), describing the position in row i and column j. The dimension of this matrix is SxS, i.e. the number of

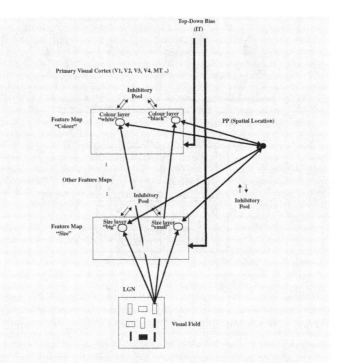

Fig. 3. Extended cortical architecture for visual attention and binding of multiple feature components (shape, colour, etc.)

items in the display is S^2. Information is processed across the different spatial locations in parallel. The authors assume that selective attention results from independent competition mechanisms operating within each feature dimension. Each visual item can be defined by M features. Each feature m can adopt $N(m)$ values, for example the feature *colour* can have the values *black* or *white* (in this case $N(colour) = 2$). For each feature map m, there are $N(m)$ layers of neurons characterizing the presence of each feature value. A cell assembly consisting of a population of fully connected excitatory integrate-and-fire spiking neurons is allocated to every location in each layer, in order to encode the presence of a specific feature value (e.g. colour *white*) at the corresponding position. The feature maps are topographically ordered; i.e. the receptive fields of the neurons belonging to cell assembly ij in one of these maps are limitated to the location ij in the retinal input. It is also assumed that the cell assemblies in layers corresponding to one feature dimension are mutually inhibitory. The posterior parietal module is bidirectionally coupled with the different feature maps and serves to bind the different feature dimensions at each item location, in order to implement local conjunction detectors. The inferotemporal connections provide top-down information, consisting of the feature values for each feature dimension of the target item. This information is fed into the system by including an extra excitatory input to the corresponding feature layer. For example, if the target is defined as small, vertical and black, then all excitatory pools at each location in the layer coding *small* in the feature map dimension *size*, in the layer coding *vertical* in the feature map *orientation* and in the layer coding *black* in the feature map *colour*, receive an extra excitatory input from the IT module.

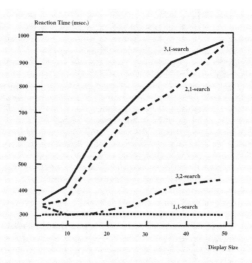

Fig. 4. Search times for feature and conjunction searches obtained utilizing the extended computational cortical model

In Fig. 4, the computational results obtained by Deco and Zihl [10] for 1,1; 2,1; 3,1 and 3,2-searches are presented. The items are defined by three feature dimensions ($M = 3$, e.g. size, orientation and colour), each having two values ($N(m) = 2$ for $m = 1,2,3$, e.g. size: big/small, orientation: horizontal/vertical, colour: white/black). For each display size, the experiment is repeated 100 times, each time with different randomly generated targets at random positions and randomly generated distractors. The mean value T of the 100 simulated search times is plotted as a function of the display size S. The slopes for all simulations are consistent with existing experimental results [21]

4   A Unified Model of Attention and Working Memory

There is much evidence that the prefrontal cortex is involved in at least some types of working memory and related processes such as planning [15]. Working memory refers to an active system for maintaining and manipulating information in mind, held during a short period, usually of seconds. Recently, Assad *et al.* [1] investigated the functions of the prefrontal cortex in working memory by analyzing neuronal activity when the monkey performs two different working memory tasks using the same stimuli and responses. In a *conditional object-response (associative) task* with a delay, the monkey was shown one of two stimulus objects

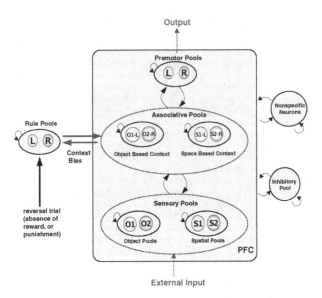

Fig. 5. Network architecture of the prefrontal cortex unified model of attention, working memory, and decision making

(O1 or O2), and after a delay had to make either a rightward or leftward ocu-lomotor saccade response depending on which stimulus was shown. In a *delayed spatial response task* the same stimuli were used, but the rule required was differ-ent, namely to respond after the delay towards the Left or Right location where the stimulus object had been shown [1]. The main motivation for such studies was the fact that for real-world behavior, the mapping between a stimulus and a response is typically more complicated than a one-to-one mapping. The same stimulus can lead to different behaviors depending on the situation, or the same behavior may be elicited by different cueing stimuli. In the performance of these tasks populations of neurons were found that respond in the delay period to the stimulus object or its position ('sensory pools'), to combinations of the response and the stimulus object or position ('intermediate pools'), and to the response required (left or right) ('premotor pools'). Moreover, the particular interme-diate pool neurons that were active depended on the task, with neurons that responded to combinations of the stimulus object and response active when the mapping was from object to behavioural response, and neurons that responded to combinations of stimulus position and response when the mapping rule was from stimulus position to response [1]. In that different sets of intermediate pop-ulation neurons are responsive depending on the task to be performed, PFC neurons provide a neural substrate for responding appropriately on the basis of an abstract rule or context.

Neurodynamics helps to understand the underlying mechanisms that imple-ment this rule-dependent mapping from the stimulus object or the stimulus position to a delayed behavioural response. Deco and Rolls [6] formulated a neu-

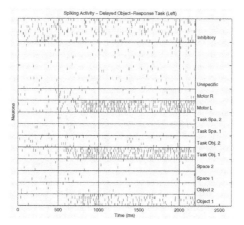

Fig. 6. Delayed Object-Response simulations. Rastergrams of randomly selected neurons for each pool in the PFC network (5 for each sensory, intermediate and premotor pool, 20 for the nonselective excitatory pool, and 10 for the inhibitory pool) after the experimental paradigms of Assad et al. (2000)

rodynamical model that builds on the integrate-and-fire attractor network by introducing a hierarchically organized set of different attractor network pools in the lateral prefrontal cortex. The hierarchical structure is organized within the general framework of the biased competition model of attention [3, 24]. There are different populations or pools of neurons in the prefrontal cortical network, as shown in Fig. 5. There are four types of excitatory pool, namely: sensory, intermediate, premotor, and nonselective. The sensory pools encode information about objects (O1 and O2), or spatial location (S1 and S2). The premotor pools encode the motor response (in our case the leftward (L) or rightward (R) oculomotor saccade). The intermediate pools are task-specific or rule-specific and perform the mapping between the sensory stimuli and the required motor response. The intermediate pools respond to combinations of the sensory stimuli and the response required, with one pool for each of the four possible stimulus-response combinations (O1-L, O2-R, S1-L and S2-R). The intermediate pools can be considered as being in two groups, one for the delayed object-response associative task, and the other for the delayed spatial response task. The intermediate pools receive an external biasing context or rule input (see Fig. 5) that reflects the current rule. The remaining excitatory neurons do not have specific sensory, response or biasing inputs, and are in a nonselective pool. All the inhibitory neurons are clustered into a common inhibitory pool, so that there is global competition throughout the network.

Figure 6 and 7 plots the rastergrams of randomly selected neurons for each pool in the network. The spatio-temporal spiking activity shows that during the short term memory delay period only the relevant sensory cue, associated future oculomotor response, and intermediate neurons maintain persistent

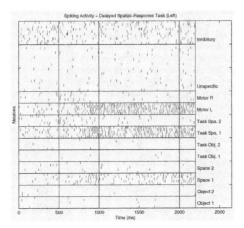

Fig. 7. Same as Fig. 6 but for the Delayed Spatial-Response simulations after the experimental paradigms of Assad et al. (2000)

activity, and build up a stable global attractor in the network. The underlying biased competition mechanisms that operate as a result of the rule/context input (see Fig. 5) biasing the correct intermediate pool neurons are very explicit in this experiment. Note that neurons in pools for the irrelevant input sensory dimension (location for the object-response associative task, and object for the delayed spatial response task), are inhibited during the sensory cue period and are not sustained during the delay short term memory period. Only the relevant single pool attractors, given the rule context, that are suitable for the cue-response mapping survive the competition and are persistently maintained with high firing activity during the short-term memory delay period. This model thus shows how a rule or context input can influence decision-making by biasing competition in a hierarchical network that thus implements a flexible mapping from input stimuli to motor outputs [24, 6]. The model also shows how the same network can implement a short term memory, and indeed how the competition required for the biased competition selection process can be implemented in an attractor network which itself requires inhibition implemented through the inhibitory neurons.

5   The Completion of the Cognitive Circuit: Reward Reversal

The prefrontal architecture shown in Fig. 5 uses a context or rule bias to influence the mapping from stimulus to response, and thus to influence decision-making. The question we now address is how the change in the rewards being received when the task is changed could implement a switch from one rule to another. Deco and Rolls [7] proposed that the rule is held in a separate attractor network as shown in Fig. 5 which is the source of the biased competition. Different sets of neurons in this network are active according to which rule is current. Consistent with this Wallis *et al.* [30] have described neurons in the primate PFC that reflect the explicit coding of abstract rules. We propose that the synapses in the recurrent attractor rule network show some adaptation, and that when an error signal is decoded by neurons in the orbitofrontal cortex (a part of the prefrontal cortex) that respond when an expected reward is not obtained [26], activation of the inhibitory interneurons reduces activity in the attractor networks, including that which holds the current rule. This reduction of firing is sufficient in the simulations to stop the dynamical interactions between the neurons that maintain the current attractor. When the attractor network starts up again after the inhibition, it starts with the other attractor state active, as the synapses of this other attractor have not adapted as a result of previous activity.

The results of this dynamical simulation demonstrate one way in which punishers, or the absence of expected rewards, can produce rapid, one trial reversal, and thus close the cognitive loop in such a way that cognitive functions are guided by the rewards being received. The rapid reversal is in this case the reversal of a rule-biased mapping from stimulus to response as illustrated in Fig. 5, but we postulate that similar rule-based networks that can be reset by non-reward underlie the rapid reversal of stimulus-reward association learning, which is a fundamental aspect of the underlying neural basis of emotion [23].

6   Conclusions

The aim of this article was to review a computational neuroscience theoretical framework that unifies microscopic, mesoscopic and macroscopic mechanisms involved in the brain functions, allowing the description of the existing experimental data (and the prediction of new results as well) at all neuroscience levels (psychophysics, functional brain imaging and single neural cells measurements). Analysis of networks of neurons each implemented at the integrate-and-fire neuronal level, and including non-linearities, enables many aspects of brain function, from the spiking activity of single neurons and the effects of pharmacological agents on synaptic currents, through fMRI and neuropsychological findings, to be integrated, and many aspects of cognitive function including visual attention, working memory and the control of behaviour by reward mechanisms to be modelled and understood. The theoretical approach described makes explicit predictions at all these levels which can be tested to develop and refine our understanding of the underlying processing, and its dynamics. We believe that this kind of analysis is fundamental for a deep understanding in neuroscience of how the brain performs complex tasks.

7   Appendix: The Mathematics of Neurodynamics

We assume that a proper level of description at the microscopic level is captured by the spiking and synaptic dynamics of one-compartment, point-like models of neurons, such as *Integrate-and-Fire-Models*. The realistic dynamics allows the use of realistic biophysical constants (like conductances, delays, etc.) in a thorough study of the realistic time scales and firing rates involved in the evolution of the neural activity underlying cognitive processes, for comparison with experimental data. We believe that it is essential of a *biologically plausible* model that the different time scales involved are properly described, because the system that we are describing is a dynamical system that is sensitive to the underlying different spiking and synaptic time courses, and the non-linearities involved in these processes. For this reason, it is convenient to include a thorough description of the different time courses of the synaptic activity, by including fast and slow excitatory receptors (AMPA and NMDA) and GABA-inhibitory receptors. A second reason why this temporally realistic and detailed level of description of synaptic activity is required, is the goal to perform realistic fMRI-simulations. These involve the realistic calculation of BOLD-signals that are intrinsically linked with the synaptic dynamics, as recently found [20]. A third reason is that one can consider the influence of neurotransmitters and pharmacological manipulations, e.g. the influence of dopamine on the NMDA and GABA receptor dynamics [31, 19], to study the effect on the global dynamics and on the related cortical functions (e.g. working memory [8, 6]). A fourth reason for analysis at the level of spiking neurons is that the computational units of the brain are the neurons, in the sense that they transform a large set of inputs received from different neurons into an output spike train, that this is the single output signal of the neuron which is connected to other neurons, and that this is therefore the level at which the information is being transferred between the neurons, and thus at which the brain's representations and computations can be understood [25, 24].

For all these reasons, the non-stationary temporal evolution of the spiking dynamics is addressed by describing each neuron by an integrate-and-fire model. The subthreshold membrane potential $V(t)$ of each neuron evolves according to the following equation:

$$C_m \frac{dV(t)}{dt} = -g_m(V(t) - V_L) - I_{syn}(t) \tag{1}$$

where $I_{syn}(t)$ is the total synaptic current flow into the cell, V_L is the resting potential, C_m is the membrane capacitance, and g_m is the membrane conductance. When the membrane potential $V(t)$ reaches the threshold θ a spike is generated, and the membrane potential is reset to V_{reset}. The neuron is unable to spike during the first τ_{ref} which is the absolute refractory period.

The total synaptic current is given by the sum of glutamatergic excitatory components NMDA and AMPA (I_{AMPA}, I_{NMDA}), and inhibitory components GABA (I_{GABA}). The total synaptic current is therefore given by:

$$I_{syn}(t) = I_{AMPA}(t) + I_{NMDA}(t) + I_{GABA}(t) \tag{2}$$

where the current generated by each receptor type follows the general form:

$$I(t) = g(V(t) - V_E) \sum_{j=1}^{N} w_j s_j(t) \tag{3}$$

and $V_E = 0$ mV for the excitatory (AMPA and NMDA) synapses and -70 mV for the inhibitory (GABA) synapses. The synaptic strengths w_j are specified by the architecture. The time course of the current flow through each synapse is dynamically updated to describe its decay by altering the fractions of open channels s according to equations with the general form:

$$\frac{ds_j(t)}{dt} = -\frac{s_j(t)}{\tau} + \sum_k \delta(t - t_j^k) \tag{4}$$

where the sums over k represent a sum over spikes emitted by presynaptic neuron j at time t_j^k, and τ is set to the time constant for the relevant receptor. In the case of the NMDA receptor, the rise time as well as the decay time is dynamically modelled, as it is slower. Details are provided by [6].

The problem now is how to analyze the dynamics and how to set the parameters which are not biologically constrained by experimentally determined values. The standard trick is to simplify the dynamics via the *mean field* approach at least for the stationary conditions, i.e. for periods after the dynamical transients, and to analyze there exhaustively the bifurcation diagrams of the dynamics. This enables a posteriori selection of the parameter region which shows in the bifurcation diagram the emergent behaviour that we are looking for (e.g. sustained delay activity, biased competition, etc.). After that, with this set of parameters, we perform the full non-stationary simulations using the *true dynamics* only described by the full integrate-and-fire scheme. The mean field study assures us that this dynamics will converge to a stationary attractor that is consistent with what we were looking for [11, 2].

In the standard mean field approach, the network is partitioned into populations of neurons, which share the same statistical properties of the afferent currents, and fire spikes independently at the same rate. The essence of the mean-field approximation is to simplify the integrate-and-fire equations by replacing after the diffusion approximation [29], the sums of the synaptic components by the average DC component and a fluctuation term. The stationary dynamics of each population can be described by the *population transfer function $F()$*, which provides the average population rate as a function of the average input current. The set of stationary, self-reproducing rates ν_i for the different populations i in the network can be found by solving a set of coupled self-consistency equations:

$$\nu_i = F(\mu_i(\nu_1, ..., \nu_N), \sigma_i(\nu_1, ..., \nu_N)) \tag{5}$$

where $\mu_i()$ and $\sigma_i()$ are the mean and standard deviation of the corresponding input current, respectively. To solve these equations, a set of first-order differential equations, describing a *fake dynamics* of the system, whose fixed point solutions correspond to the solutions of Eq. 5, is used :

$$\tau_i \frac{d\nu_i(t)}{dt} = -\nu_i(t) + F(\mu_i(\nu_1, ..., \nu_N), \sigma_i(\nu_1, ..., \nu_N)) \tag{6}$$

The standard mean field approach neglects the temporal properties of the synapses, i.e. considers only delta-like spiking input currents. Consequently, after this simplification, the transfer function $F()$ is an Ornstein-Uhlenbeck solution for the simplified integrate-and-fire equation $\tau_x \frac{dV(t)}{dt} = -V(t) + \mu_x + \sigma_x \sqrt{\tau_x} \eta(t)$, as detailed by [2]. An extended mean-field framework, which is consistent with the integrate-and-fire and synaptic equations described above, i.e. that considers both the fast and slow glutamatergic excitatory synaptic dynamics (AMPA and NMDA), and the dynamics of GABA-inhibitory synapses, was derived by [2].

References

1. W. F. Asaad, G. Rainer, and E. K. Miller. Task-specific neural activity in the primate prefrontal cortex. *Journal of Neurophysiology*, 84:451–459, 2000.
2. N. Brunel and X. Wang. Effects of neuromodulation in a cortical networks model of object working memory dominated by recurrent inhibition. *Journal of Computational Neuroscience*, 11:63–85, 2001.
3. L. Chelazzi. Serial attention mechanisms in visual search: a critical look at the evidence. *Psychological Research*, 62:195–219, 1998.
4. S. Corchs and G. Deco. Large-scale neural model for visual attention: integration of experimental single cell and fMRI data. *Cerebral Cortex*, 12:339–348, 2002.
5. G. Deco and T. S. Lee. A unified model of spatial and object attention based on inter-cortical biased competition. *Neurocomputing*, 44–46:775–781, 2002.
6. G. Deco and E. T. Rolls. Attention and working memory: a dynamical model of neuronal activity in the prefrontal cortex. *European Journal of Neuroscience*, 18:2374–2390, 2003.
7. G. Deco and E. T. Rolls. Synaptic and spiking dynamics underlying reward reversal in the orbitofrontal cortex. *Journal of Cognitive Neuroscience*, page in press, 2004.
8. G. Deco, E. T. Rolls, and B. Horwitz. 'What' and 'where' in visual working memory: a computational neurodynamical perspective for integrating fmri and single-neuron data. *Journal of Cognitive Neuroscience, in press*, 2003.
9. G. Deco and J. Zihl. A neural model of binding and selective attention for visual search. In D. Heinke, G. Humphreys, and A. Olson, editors, *Connectionist Models in Cognitive Neuroscience – The 5th Neural Computation and Psychology Workshop*, pages 262–271. Springer, Berlin, 1999.
10. G. Deco and J. Zihl. Top-down selective visual attention: a neurodynamical approach. *Visual Cognition*, 8:119–140, 2001.
11. P. Del Giudice, S. Fusi, and M. Mattia. Modeling the formation of working memory with networks of integrate-and-fire neurons connected by plastic synapses. *Journal of Physiology Paris*, page in press, 2003.
12. R. Desimone and J. Duncan. Neural mechanisms of selective visual attention. *Annual Review of Neuroscience*, 18:193–222, 1995.
13. J. Duncan. Cooperating brain systems in selective perception and action. In T. Inui and J. L. McClelland, editors, *Attention and Performance XVI*, pages 549–578. MIT Press, Cambridge, MA, 1996.
14. J. Duncan and G. Humphreys. Visual search and stimulus similarity. *Psychological Review*, 96:433–458, 1989.
15. J. Fuster. Executive frontal functions. *Experimental Brain Research*, 133:66–70, 2000.
16. F. Hamker. The role of feedback connections in task-driven visual search. In D. Heinke, G. Humphreys, and A. Olson, editors, *Connectionist Models in Cognitive Neuroscience – The 5th Neural Computation and Psychology Workshop*, pages 252–261. Springer, Berlin, 1999.

17. S. Kastner, P. De Weerd, R. Desimone, and L. Ungerleider. Mechanisms of directed attention in the human extrastriate cortex as revealed by functional MRI. *Science*, 282:108–111, 1998.

18. S. Kastner, M. Pinsk, P. De Weerd, R. Desimone, and L. Ungerleider. Increased activity in human visual cortex during directed attention in the absence of visual stimulation. *Neuron*, 22:751–761, 1999.

19. D. Law-Tho, J. Hirsch, and F. Crepel. Dopamine modulation of synaptic transmission in rat prefrontal cortex: An in vitro electrophysiological study. *Neuroscience Research*, 21:151–160, 1994.

20. N. K. Logothetis, J. Pauls, M. Augath, T. Trinath, and A. Oeltermann. Neurophysiological investigation of the basis of the fMRI signal. *Nature*, 412:150–157, 2001.

21. P. T. Quinlan and G. W. Humphreys. Visual search for targets defined by combination of color, shape, and size: An examination of the task constraints on feature and conjunction searches. *Perception and Psychophysics*, 41:455–472, 1987.

22. J. Reynolds, L. Chelazzi, and R. Desimone. Competitive mechanisms subserve attention in macaque areas V2 and V4. *Journal of Neuroscience*, 19:1736–1753, 1999.

23. E. T. Rolls. *The Brain and Emotion*. Oxford University Press, Oxford, 1999.

24. E. T. Rolls and G. Deco. *Computational Neuroscience of Vision*. Oxford University Press, Oxford, 2002.

25. E. T. Rolls and A. Treves. *Neural Networks and Brain Function*. Oxford University Press, Oxford, 1998.

26. S. J. Thorpe, E. T. Rolls, and S. Maddison. Neuronal activity in the orbitofrontal cortex of the behaving monkey. *Experimental Brain Research*, 49:93–115, 1983.

27. A. Treisman. Perceptual grouping and attention in visual search for features and for objects. *Journal of Experimental Psychology: Human Perception and Performance*, 8:194–214, 1982.

28. A. Treisman. Features and objects: The fourteenth Barlett memorial lecture. *The Quarterly Journal of Experimental Psychology*, 40A:201–237, 1988.

29. H. Tuckwell. *Introduction to theoretical neurobiology*. Cambridge University Press, Cambridge, 1988.

30. J. Wallis, K. Anderson, and E. Miller. Single neurons in prefrontal cortex encode abstract rules. *Nature*, 411:953–956, 2001.

31. P. Zheng, X.-X. Zhang, B. S. Bunney, and W.-X. Shi. Opposite modulation of cortical N-methyl-D-aspartate receptor-mediated responses by low and high concentrations of dopamine. *Neuroscience*, 91:527–535, 1999.

Modeling Attention:
From Computational Neuroscience to Computer Vision

Fred H. Hamker

[1] Allgemeine Psychologie, Psychologisches Institut II
Westf. Wilhelms-Universität,
48149 Münster, Germany
fhamker@uni-muenster.de
http://wwwpsy.uni-muenster.de/inst2/lappe/Fred/FredHamker.html
[2] California Institute of Technology,
Division of Biology 139-74,
Pasadena, CA 91125, USA

Abstract. We present an approach to modeling attention which originates in computational neuroscience. We aim at elaborating the underlying mechanisms of attention by fitting the model with data from electrophysiology. Our strategy is to
either confirm, reject, modify or extend the model to accumulate knowledge in a
single model across various experiments. Here, we demonstrate the present state
of the art and show that the model allows for a goal-directed search for an object
in natural scenes.

1 Introduction

Visual Search and other experimental approaches have demonstrated that attention plays
a crucial role in human perception. Understanding attention and human vision in general
could be beneficial to computer vision, especially in vision tasks that are not limited to
specific and constrained environments. We discuss recent findings and hypotheses in
the neurosciences that have been modeled by approaches from computational neuroscience. Neuroscience gives an insight into the brain which allows to further constrain
algorithms of attention. In computational neuroscience the topics of interest are usually
focused on a specific mechanism and networks often comprise only a relatively low
number of cells and artificial inputs are used. Thus, scaleability becomes an important
issue. A transfer of knowledge from computational neuroscience to computer vision
requires at least the solution of three constraints: i) Does the number of cells influence
the convergence of the algorithm? ii) Can the preconditions of the proposed solution
be embedded into the systems level? iii) Can the model be demonstrated to operate on
natural scenes?

In this contribution we derive a computational principle that allows to model large
scale systems and vision in natural scenes. We demonstrate an approach for object
detection in natural scenes. We suggest that goal directed attention and object detection
are necessarily coupled, since an efficient deployment of attention benefits from an at
least partial match of the encoded objects with the target.

L. Paletta et al. (Eds.): WAPCV 2004, LNCS 3368, pp. 118–132, 2005.

2   Spatial Attention

2.1   Gain Control

Single cell recordings have revealed that the neural response is enhanced in a multiplicative fashion when attention is directed to a single stimulus location [23], [15]. The neural correlate of such a multiplicative effect is still under discussion. It has been suggested that the gain of a neuronal response to excitatory drive is decreased by increasing the level of both, excitatory and inhibitory, background firing rates in a balanced manner [1]. On a more abstract level a feedback signal could increase the gain of the feedforward pathway in a multiplicative fashion [18], [9], [22]. We investigated such a gain control mechanism by simulating a V4 layer which receives input from a V2 population. We consider feedforward, lateral excitory and inhibitory input and spatial bias.

Given a neural population in V4 and a feedforward input $I^\uparrow$ we have proposed that a cell's response over time $r(t)$ can be computed by a differential equation:

$$\tau \frac{d}{dt} r_k(t) = I_k^\uparrow + I_k^N + I_k^A - I_k^{inh} \tag{1}$$

Inhibition I_k^{inh} introduces competition among cells and normalizes the cell's response by a shunting term. I_k^N describes the lateral influence of other cells in the population. Spatial attention is proposed to emerge from the modulation of the feedforward signal by feedback A_x prior to spatial pooling:

$$I_k^A = f(w_{i,x} a_{i,x}^A); \quad a_{i,x}^A = I_{i,x}^\uparrow \cdot A_x \quad f = \max_{i,x} \tag{2}$$

We presented a stimulus to a population of 11 orientation selective cells for 150 ms and computed the average activity of each cell with and without a spatial bias. Consistent with the findings, we observed that the response on the population level is close to a multiplicative increase of the gain (Fig. 1). If we consider neural cells as feature detectors indicating the probability that the encoded feature is present in the scene, the function of gain control is increasing the probability of a feature being detected.

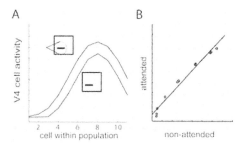

Fig. 1. Gain control. (A) Population responses to a single horizontal bar 100 ms after stimulus onset with and without a spatial bias. Each cell encodes a different orientation. (B) The firing rate of each cell in the non-attended case is plotted against the attended case (see [15]). The model shows approximately a multiplicative gain increase (CorrCoeff=0.99) of 13% (slope=1.13)

2.2   Contrast Dependence

A simple multiplicative gain control model (response gain model) predicts that the effect of attention increases with stimulus contrast. However, this is not a very useful strategy, since a high contrast stimulus is already salient. If we assume at least some parallel processing, a too high gain to an already salient stimulus could suppress other potentially relevant responses. Indeed the brain uses a strategy in which the magnitude of the attentional modulation decreases with increasing contrast [21], [13]. Attention results rather in a shift of the contrast response function (contrast gain model). This was experimentally tested by presenting a single luminance-modulated grating within the receptive field of a V4 cell. The monkey was then instructed to either attend towards the stimulus location or towards a location far outside of the receptive field. An increase of the stimulus contrast resulted in an increase of the neural response and in a decrease of the difference between both conditions (Fig. 2A).

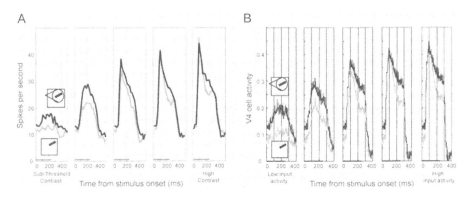

Fig. 2. Contrast dependent effect of attention. (A) Averaged single cell responses over time, with increasing contrast. The data was provided by J. Reynolds [21]. (B) Simulation results of our model. The initial burst at high contrast levels occurs due do the delayed inhibitory response I^{inh}. Similar to the data, the timing of the attention effect shifts with increasing contrast from early to late

In the model of Reynolds and Desimone [19] spatial attention affects the weight of feedforward excitation and feedforward inhibition. As the activation of the input increases the inhibition increases as well and the cell's response will saturate at a level where excitation and inhibition are balanced. However, a re-implementation of this model shows that the model replicates their finding on the level of the mean response over the whole presentation time, but it does not account very well for the observed temporal course of activity and the timing of the attention effect [22]. The model of Spratling and Johnson [22] accounts better for the temporal course of activity, but different as indicated by the data, the decrease in the magnitude of the attentional modulation occurs only at high contrast. A potential problem of this model is, that it explains the decrease of the attentional modulation by a saturation effect, which occurs only at high activity.

We postulate that the efficiency of the feedback signal decreases with increasing strength of the cell. Thus, the feedback signal A_x (eq. 2) which determines the gain factor $1 + A_x$ is combined with an efficiency term using the activity of the output cell k.

$$A_{k,x} = \sigma(\alpha - r_k) \cdot A_x \qquad (3)$$

with $\sigma(a) = max(a, 0)$. We applied the extended model to simulate the effect of contrast dependence. Increasing contrast was simulated by increasing the stimulus strength. The model accounts quite well for the findings, even in the temporal course of activity (Fig. 2B). The magnitude of the attention effect is not explained by the saturation of the cell. Thus, the contrast dependency of attention is consistent with an effective modulation of the input gain by the activity of the cell. An answer towards the underlying exact neural correlate, however, requires more research.

To further demonstrate that the model is consistent with the contrast gain model, the magnitude of attention on the time averaged response with varying stimulus strength is shown (Fig. 3). We computed the mean response beginning from stimulus onset (Fig. 3A) and the mean over the first initial response (Fig. 3B) to show the timing of attention. For high contrast stimuli attention is most prominent in the late response and almost diminishes in the early response. Please note, the spatial feedback signal itself is constant.

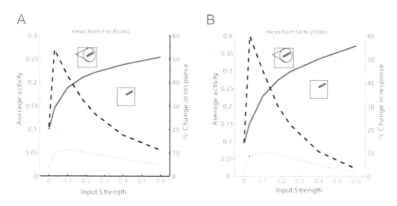

Fig. 3. Simulation of the attention effect on the mean response by varying the strength of the input stimulus. The dotted line shows the absolute difference between the attended and non-attended condition, and the dashed line the difference in percent. Consistent with the contrast gain model [21], the primary effect of attention occurs with low input activty. (A) Time average over the whole neural response after stimulus onset. (B) Time average over the initial burst

2.3   Biased Competition

It has been observed that neuronal populations compete with each other when more than a single stimulus are presented within a receptive field. Such competition can be biased by top-down signals [4]. As a result, the irrelevant stimulus is suppressed as if only the attended one had been presented. Numerous experiments have supported this framework.

Reynolds, Chelazzi and Desimone observed competitive interactions by placing two stimuli (reference and probe) within the receptive field of a V4 neuron [20]. They found that when spatial attention was directed away from the receptive field, the response to both stimuli was a weighted average of the responses to the stimuli presented in isolation. If the reference elicits a high firing rate and the probe a low firing rate, then the response to both is in between. Attending to the location of one of the stimuli biases the competition towards the attended stimulus.

We modeled this experiment by presenting now two stimuli to our neural population. In the attended case the gain of the input from one location is increased. The simulation results of our population approach fit with the experimental data (Fig. 4). A slope of 0.5 indicates that reference and probe are equally well represented by the population. The small positive y-intercept signifies a slight overall increase in activity when presenting a second stimulus along with the first. Attending to the probe increases the slope (not shown), indicating the greater influence of the attended probe over the population. Attending to the reference reduces the slope, signifying the greater influence of the attended reference stimulus. Attention in general enhances the overall response within the population, which is observed by the greater upward shift of the sensory interaction index as compared to the attend away condition.

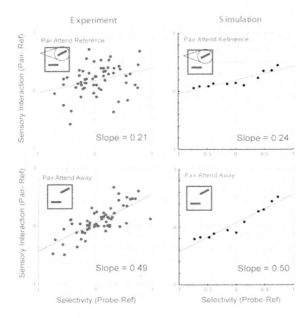

Fig. 4. Comparison of the simulation results with the experimental data (modified from [20]) to investigate the influence of attention on the sensory interaction. For each cell, its selectivity index is plotted over its sensory interaction. A selectivity value of 0 indicates identical responses to reference and probe in isolation, a positive value a preference towards the probe and a negative a preference towards the reference. An interaction index of 0 signifies that the cell is unaffected by adding a probe. Positive values indicate that the cell's response to the reference is increased by adding a probe and negative values signify a suppression by the probe

Along with the experimental data, Reynolds, Chelazzi and Desimone [20] demonstrated that a feedforward shunting model [5] can account for their findings. In their model competition among cells occurs due to feedforward inhibition from V2 cells onto V4 cells. In our approach competition occurs after pooling. It is based on lateral short range excitory and long range inhibitory connections within the population, which is in accordance with findings in V4. Other models [6] [3] [22] have referred to the findings of Reynolds, Chelazzi and Desimone [20] as well, but no quantitative comparison with the experimental data (Fig. 4) has been given.

3   Feature-Based Attention

3.1   Feature-Similarity

With reference to feature-based attention Treue and Martínez Trujillo [24] have proposed the Feature-Similarity Theory of attention. Their single cell recordings in area MT revealed that directing attention to a feature influences the encoding of a stimulus even when the second stimulus is presented outside of the receptive field. They proposed that attending towards a feature could provide a global feedback signal which affects other locations than the attended one as well. Feedback can be a very useful mechanism for a feature-based selection, as already demonstrated in early computational models of attention [25].

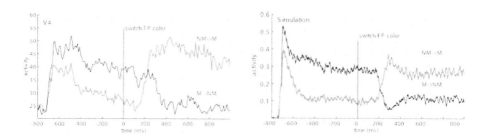

Fig. 5. Temporal course of activity in the match and non-match condition of V4 cells and simulated cells. The scene is presented at -800 ms. Cells representing the potential target object show an enhanced activity. If the fixation point color switches to another color at 0 ms, the activity follows the definition of the target. Neurons previously representing potential targets change into distractors and vice versa. The left figure shows the original data provided by B. Motter (the published data [17] does not show the activity after stimulus onset)

In an earlier experiment that presumably revealed feature-based attention effects the knowledge of a target feature increased the activity of V4 cells [17]. The task required to report the orientation of an item that matches the color of the fixation point. Since the display during the stimulus presentation period contains several possible targets, the monkey had to wait until the display contained only one target. Even during this stimulus presentation period, V4 neurons showed an enhanced activity if the presented colour or luminance items matched the target (Fig. 5 left). This dynamic effect is thought to occur in parallel across the visual field, segmenting the scene into possible candidates and background.

We simulated this experiment by presenting input to six $x \in \{1 \ldots 6\}$ V4 populations containing 11 cells i in each dimension $d \in \{color, orientation\}$ (eq. 4).

$$\tau \frac{d}{dt} r^{V4}_{d,i,x} = I^{\uparrow}_{d,i,x} + I^{N}_{d,i,x} + I^{A}_{d,i,x} - I^{inh}_{d,x} \qquad (4)$$

We also model one IT population, whose receptive field covers all V4 receptive fields (Fig. 6). Let us assume the model is supposed to look for red items. This is implemented by generating a population of active prefrontal cells (PF) representing a red target template. At $t = -900ms$ we activate the target template in PF and present the inputs at $t = -800ms$. The input activity travels up from V4 to IT. Once the activity from V4 enters IT, competition gets biased by feedback from prefrontal cells. They in turn project back to V4 and enhance the gain of the V4 input. Thus, the term $I^{A}_{d,i,x}$ is a result of the bottom-up signal $I^{\uparrow}_{d,i,x}$ modulated by the feedback signal $r^{IT}_{d,j}$ with $w^{ITt,V4}_{i,j}$ as the strength of the feedback connection:

$$I^{A}_{d,i,x} = f \left(I^{\uparrow}_{d,i,x} \sigma(\alpha - r^{V4}_{d,i,x}) \cdot \max_{j}(w^{ITt,V4}_{i,j} \cdot r^{IT}_{d,j}) \right) \qquad (5)$$

Feedback in the "object pathway" operates feature specific and largely location unspecific. By changing the target template at $t = 150ms$ the model now switches into a state where again all items of the target color in V4 are represented by a higher firing rate than those with a non-matching color (Fig. 5 right). Consistent with the Feature-Similarity Theory, the enhancement of the gain depends on the similarity of the input population with the feedback population.

The computationally challenging task of this experiment is to enable the model to switch its internal representation. Our gain control mechanism supports rapid switches, because feedback acts on the excitatory input of a cell and not on its output activity. Due to the switch in the PF activity, the population in IT encoding red looses its feedback signal whereas the one for green receives support. As a result, the prioritized encoding in IT changes, and the whole system switches to a state where populations encoding the new target feature are represented by a higher activity.

4   Attention on the Systems Level

A top-down feature-specific signal has also been revealed in IT cells during visual search [2]. In this experiment an object was presented to a monkey, which after a brief delay, had to be detected in a visual search scene. The monkey was trained to indicate the detection by shifting its gaze from the fixation point towards the target. Chelazzi found that the initial activation of IT neurons is largely stimulus driven and cells encoding target and non-target become activated. Since different populations compete for representation, typically the cells encoding the non-target get suppressed. A computational approach by Usher and Niebur [26] has shown that a parallel competition based on lateral interactions and a top-down bias is sufficient to qualitatively replicate some of those

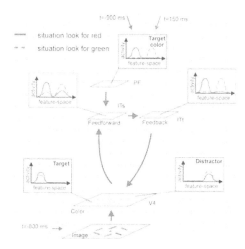

Fig. 6. Illustration of the pathway for "object recognition". At each V4 receptive field we model an arbitrary color space. We only show one target and one distractor. Due to the gain control by feedback the population encoding the target color gets enhanced and the one for the distractor is suppressed. The network settles in a state where all items of the target color in V4 are represented by a higher firing rate than those with a non-matching color. The second situation is "looking for green items", indicated by the dashed activity curves

findings. However, we have argued that the planning of an eye movement towards the target should produce a spatial reentry signal directed to the target location [8]. This prediction recently received further evidence by a study in which a microstimulation in the frontal eye field resulted in a modulation of the gain in V4 cells [16].

We have modeled the visual search experiment on the systems level by a model consisting of areas V4, IT, FEF and PFC (Fig. 7A,B). Thus, spatial and feature-based attention are now brought together in a single model. V4 cells receive a top-down signal from IT and the FEF, which both add up:

$$I_{d,i,x}^A = f\left(I_{d,i,x}^\uparrow \cdot \sigma() \cdot \max_j w_{i,j}^{\text{ITt,V4}} \cdot r_{d,j}^{\text{IT}} \right) + f\left(I_{d,i,x}^\uparrow \cdot \sigma() w^{\text{FEFm,V4}} r_x^{\text{FEFm}} \right) \quad (6)$$

with $\sigma() = \sigma(\alpha - y_{d,k,x}^{V4})$ and $w^{\text{FEFm,V4}}$ defines the weight of the feedback from the FEF. For implementation details please refer to [8].

Our simulation result matches even the temporal course of activity of the experimental data (Fig. 7C). The model predicts that the firing rate of V4 and IT cells show an early feature-based effect and a late spatial selectivity (after 120 ms). In the 'Target Absent' condition where the cue stimulus is different from the stimuli in the choice array no spatial reentry signal emerges since in this case a saccade has to be withheld. The model does not contain any control units or specific maps that implement attention. The proposed gain control and competition allows higher areas to influence processing in lower areas. As a result, suppressive and facilitatory effects occur, commonly referred to as "attention". Thus, attention can emerge on the network level and does not have to be explicitly implemented.

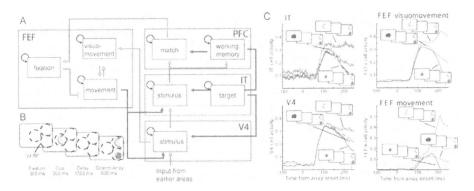

Fig. 7. (A) Sketch of the simulated areas. Each box represents a population of cells. The activation of those populations is a temporal dynamical process. Bottom-up (driving) connections are indicated by a bright arrow and top-down (modulating) connections are shown as a dark arrow. (B) Simulation of the experiment. The objects are represented by a noisy population input. RF's without an object just have noise as input. Each object is encoded within a separate RF, illustrated by the dashed circle, of V4 cells. All V4 cells are within the RF of the IT cell population. (C) Activity within the model areas aligned to the onset of the search array in the different conditions

5   Large Scale Approach for Modeling Attention

Our earlier simulations have shown that competition among feature representations could be a useful mechanism to filter out irrelevant stimuli for object recognition. A spatial focus of attention can reduce the influence of features outside the focus, whereas a competition among features could have the potential to select objects without the need of a segmentation on the image level. We now demonstrate how attention emerges in the process of detecting an object in a natural scene [7]. In extension to a mere biased competition we show that top-down signals can be modeled as an expectation, which alters the gain of the feedforward signal.

5.1   Overview

The idea is that all mechanisms act directly on the processed variables and modify their conspicuity. Each feature set is modeled as a continuous space with $i \in N$ cells at location $\mathbf{x} = (x_1, x_2)$ by assigning each cell a conspicuity $r_{d,i,\mathbf{x}}$. From the feature maps we determine contrast maps according to a measure of stimulus-driven saliency (Fig. 8). Feature and contrast maps are then combined into feature conspicuity maps which encode the feature and its initial conspicuity by means of a population code.

The conspicuity of each feature is altered by the target template. A target object is defined by the expected features $\hat{r}^F_{d,i}$. We infer the conspicuity of each feature $r_{d,i,\mathbf{x}}$ by comparing the expected features $\hat{r}^F_{d,i}$ with the bottom-up signal $r^{\uparrow}_{d,i,\mathbf{x}}$. If the bottom-up signal is similar to the expectation we increase the conspicuity. Such a mechanism enhances in parallel the conspicuity of all features at level II which are similar to the target template. We perform the same procedure on level I where the expected features are those from level II. In order to detect an object in space we combine the conspicuity across all d channels in the perceptual map and generate an expectation in space $\hat{r}^L_{\mathbf{x}}$ in the movement map. The higher the individual conspicuity $r_{d,i,\mathbf{x}}$ across d at one location relative to all other locations the higher is the expectation in space $\hat{r}^L_{\mathbf{x}}$ at this location. Thus, a location with high conspicuity in different channels d tends to have a high expectation in space $\hat{r}^L_{\mathbf{x}}$. Analogous to the inference in feature space we iteratively compare the expected location $\hat{r}^L_{\mathbf{x}}$ with the bottom-up signal $r^\uparrow_{d,i,\mathbf{x}}$ in $\mathbf{x}$ and enhance the conspicuity of all features with a similarity of expectation and bottom-up signal. The conspicuity is normalized across each map by competitive interactions. Such interative mechanisms finally lead to a preferred encoding of the features and space of interest. Thus, attention emerges by the dynamics of vision.

Preprocessing: We compute feature maps for Red-Green opponency (RG), Blue-Yellow opponency (BY), Intensity (I), Orientation (O), and Spatial Resolution (σ). We determine the initital conspicuity by center-surround operations [11] from the feature maps which gives us the contrast maps. The feature-conspicuity maps combine the feature and conspicuity into a population code, so that at each location we encode each feature and its related conspicuity.

Level I: Level I has d channels which receive input from the feature conspicuity maps: $r_{\theta,i,\mathbf{x}}$ for orientation, $r_{I,i,\mathbf{x}}$ for intensity, $r_{RG,i,\mathbf{x}}$ for red-green opponency, $r_{BY,i,\mathbf{x}}$ for blue-yellow opponency and $r_{\sigma,i,\mathbf{x}}$ for spatial frequency (Fig. 8). The expectation of features at level I originates in level II $\hat{r}^{I_F}_{d,i,\mathbf{x}'} = r^{II}_{d,i,\mathbf{x}}$ and the expected location in the movement map $\hat{r}^{I_L}_{\mathbf{x}'} = r^m_{\mathbf{x}'}$. Please note that even level II has a coarse dependency on location.

Level II: The features with their respective conspicuity and location in layer I project to layer II, but only within the same dimension d, so that the conspicuity of features at several locations in level I converges onto one location in level II. We simulate a map containing 9 populations with overlapping receptive fields. We do not increase the complexity of features from level I to level II. The expected features at level II originate in the target template $r^{II_F}_{d,i,\mathbf{x}} = w \cdot r^T_{d,i}$ and the expected location in the movement map $\hat{r}^{II_L}_{\mathbf{x}} = w \cdot r^m_{\mathbf{x}}$

Perceptual Map: The perceptual map (v) indicates salient locations by integrating the conspicuity of level I and II across all channels. In addition to the the conspicuity in level I and II we consider the match of the target template with the features encoded in level I by the product $\prod_d \max_{i,\mathbf{x}' \in RF(\mathbf{x})} r^T_{d,i} \cdot r^I_{d,i,\mathbf{x}'}$. This implements a bias to locations with a high joint probability of encoding all searched features in a certain area.

Movement Map: The projection of the perceptual map onto the movement map (m) transforms the salient locations into a few candidate locations which provide the expected location for level I and level II units. We achieve this by subtracting the average saliency from the saliency at each location $w^v r^v_{\mathbf{x}} - w^v_{inh} \sum_{\mathbf{x}} r^v_{\mathbf{x}}$. Simultaneously, the movement units indicate the target location of an eye movement.

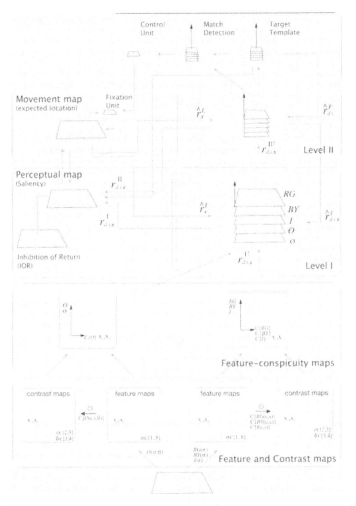

Fig. 8. Model of attentive vision. From the image we obtain 5 feature maps. For each feature at each location **x** we compute its conspicuity in the contrast maps and then combine feature and conspicuity into feature-conspicuity maps. This initial, stimulus-driven conspicuity is now dynamically updated within a hierarchy of levels. From level I to level II we pool across space to achive a representation of features with a coarse coding of location. The target template $\hat{r}^F_{d,i}$ holds the to be searched pattern regardless of its location and enhances the gain of level II cells which match the pattern of the template. $\hat{r}^F_{d,i,x}$ sends the information about relevant features further downwards to level I cells to localize objects with the relevant features. In oder to identify candidate objects the perceptual map integrates across all 5 channels to determine the saliency. The saliency is then used to compute the expected locations of an object $\hat{r}^L_{\mathbf{x}}$ in the movement map, which in turn enhances the conspicuity of all features at level I and II at these locations. Match detection cells fire, if the encoded features in level II match with the target template. This information can be used to control the fixation unit

5.2   Results

We now demonstrate the performance of our approach on an object detection task (Fig. 9). We present an object to the model for 100 ms and let it memorize some of its features as a target template. We do not give the model any hints which feature to memorize. As in the experiment done by Chelazzi, the model's task is to make an eye movement towards the target. When presenting the search scene, level II cells that match the target template quickly increase their activity to guide level I cells. In the blue-yellow channel at level I the target template is initially not dominant but the modulation by the expectation from level II overwrites the initial conspicuity. Thus, the features of the object of interest are enhanced prior to any spatial focus of attention which allows to guide the planning of the saccade in the perceptual and movement map sufficiently well. Saliency is not encoded in a single map. Given that level I cells have a spatially localized receptive field and show an enhanced response to relevant stimuli, they could be interpreted to encode a saliency map as well, which is consistent with recent findings [14]. A feature independent saliency map is achieved by the integration across all channels. The process of planning an eye movement provides a spatially organized reentry signal, which enhances the gain of all cells at the target location of the intended eye movement. Thus, spatial attention could be interpreted as a shortcut of the actual planned eye movement. Under natural viewing conditions spatial attention and eye movement selection are automatically co-ordinated such that prior to the eye movement the amount of reentry is maximized at the endpoint and minimized elsewhere. This would facilitate planning processes to evaluate the consequences of the planned action.

6   Discussion

We have modeled several attention experiments to derive the basic mechanisms of visual processing and attention. Initially we have focused on the gain control mechanism and demonstrated that an input gain model allows for a quantitative match with existing data. If we further assume that the gain factor decreases with the activity of the cell, the model is consistent with a contrast gain model. We then extended the model to simulate more complex tasks and to model the behavioral response as well. Again, we have been able to achieve a good match with the data. So far the model provides a comprehensive account of attention, specifically on the population averaged neural firing rate. Certainly attention is still more complicated than covered by the present model. However, the good fit with many existing data makes us believe the model contains at least several relevant local mechanisms that determine attention in the brain. Almost 20 years after the influential computational model of Koch and Ullman [12] was published, single cell recordings and computational modeling have now discovered a more fine graded model in which attention is explained on the systems level rather than by a selection within a single area.

In regard of this emerging new view on attention we investigated if the derived principles of the distributed nature of attention can be demonstrated to provide something useful beyond fitting experimental data. Thus, we tested an extension of the model on a goal-directed object detection task in natural scenes. We are confident that this joint approach gives the model a high potential for future computer vision tasks. The present

A

B   Level II

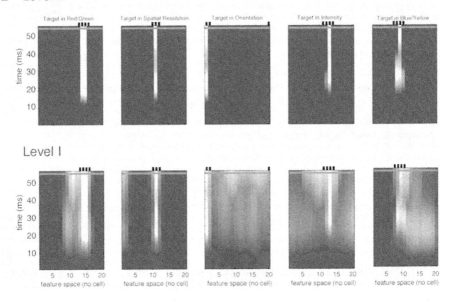

Level I

Fig. 9. Visual search in natural scenes. The asprin bottle in the upper left corner was presented to the model before the scene appeared, and in each dimension the most conspicious feature was memorized in order to generate a target template. Then the model searched for the target. A) Indication of the first eye movement, which directly selects the target. B) Conspicuity values of level II and level I cells in all channels over time. At each level the maximum response for each feature is shown, regardless of the receptive field of the cell. The strength of conspicuity is indicated by brightness. The target template is indicated by the bars at the top of each figure. The conspicuity of each feature occurs first in level I and then travels upwards to level II. Level II, however, first follows the target template, which then travels downwards to level I. This top-down inference is clearly visible in the blue-yellow channel (most right), where initially other features than the target feature are conspicious. The effect of the spatially localized reentry signal is best visible in the Intensity channel (second right). Prior to the eye movement several cells gain in activity, independent to their similarity of the encoded feature to the target template

demonstration of an object detection in natural scenes is very valuable from the viewpoint of attention, since it demonstrates that the derived principles even hold for the postulated three constraints: i) large number of cells ii) systems level and iii) natural scenes.

From the viewpoint of computer vision, we are aware that such an object detection task can be solved by classical methods. The advantage of our approach, however, lies in the integration of recognition and attention into a common framework. Attention improves object recognition, specifically in cluttered scenes, but only if attention can be properly guided to the object of interest. Feature-specific feedback within the object recognition pathway, gain control and competitive interactions directly enhance the features of interest and guide spatial attention to the object of interest. Partial attention improves further analysis which in turn helps to direct attention. We propose that the direction of attention and recognition must be an iterative process to be effective. In the present version we only used simple cues. Thus, future work has to focus on the learning of effective feedforward and feedback filters for shape recognition and object grouping.

Acknowledgements

The main part of the presented research has been done at Caltech. In this respect, I thank Christof Koch for his support and Rufin VanRullen for helpful discussions. I am grateful to John Reynolds and Brad Motter for providing data showing attention effects on V4 cells. Most of this research was supported by DFG HA2630/2-1 and in part by the ERC Program of the NSF (EEC-9402726).

References

1. Chance, F.S., Abbott, L.F., Reyes, A.D. (2002) Gain modulation from background synaptic input. Neuron, 35, 773-782.
2. Chelazzi, L., Miller, E.K., Duncan, J., Desimone, R. (1993) A neural basis for visual search in inferior temporal cortex. Nature, 363, 345-347.
3. Corchs, S., Deco, G. (2002) Large-scale neural model for visual attention: integration of experimental single-cell and fMRI data. Cereb. Cortex, 12, 339-348.
4. Desimone, R., Duncan, J. (1995) Neural mechanisms of selective attention. Anu. Rev. of Neurosc., 18, 193-222.
5. Grossberg, S. (1973) Contour enhancement short term memory, and constancies in reverberating neural networks, Studies in Applied Mathematics, 52, 217-257.
6. Grossberg, S., Raizada, R. (2000) Contrast-sensitive perceptual grouping and object-based attention in the laminar circuits of primary visual cortex. Vis. Research, 40, 1413-1432.
7. Hamker, F.H., Worcester, J. (2002) Object detection in natural scenes by feedback. In: H.H. Bülthoff et al. (Eds.), Biologically Motivated Computer Vision. Lecture Notes in Computer Science. Berlin, Heidelberg, New York: Springer Verlag, 398-407.
8. Hamker, F.H. (2003) The reentry hypothesis: linking eye movements to visual perception. Journal of Vision, 11, 808-816.
9. Hamker, F.H. (2004) Predictions of a model of spatial attention using sum- and max-pooling functions. Neurocomputing, 56C, 329-343.
10. Hamker, F.H. (2004) A dynamic model of how feature cues guide spatial attention. Vision Research, 44, 501-521.

11. Itti, L., Koch, C., Niebur, E. (1998) A model of saliency-based visual attention for rapid scene analysis. IEEE Transactions on Pattern Analysis and Machine Intelligence (PAMI), 20, 1254-1259.
12. Koch C, Ullman S (1985) Shifts in selective visual attention: towards the underlying neural circuitry. Human Psychology 4:219-227.
13. Martínez Trujillo, J.C., Treue, S., (2002) Attentional modulation strength in cortical area MT depends on stimulus contrast. Neuron, 35:365-370.
14. Mazer, J.A., Gallant ,J.L. (2003) Goal-related activity in V4 during free viewing visual search. Evidence for a ventral stream visual salience map. Neuron, 40, 1241-1250.
15. McAdams, C.J., Maunsell, J.H. (1999) Effects of attention on orientation-tuning functions of single neurons in macaque cortical area V4, J. Neurosci. 19, 431-441.
16. Moore, T., Armstrong, K.M. (2003) Selective gating of visual signals by microstimulation of frontal cortex. Nature, 421, 370-373.
17. Motter, B.C. (1994) Neural correlates of feature selective memory and pop-out in extrastriate area V4. J. Neurosci., 14, 2190-2199.
18. Nakahara, H., Wu, S., Amari, S. (2001) Attention modulation of neural tuning through peak and base rate, Neural Comput., 13, 2031-2047.
19. Reynolds, J.H., and Desimone, R. (1999) The Role of Neural Mechanisms of Attention in Solving the Binding Problem, Neuron, 24:19-29.
20. Reynolds, J.H., Chelazzi, L., Desimone R. (1999) Competetive mechanism subserve attention in macaque areas V2 and V4, J. Neurosci., 19, 1736-1753.
21. Reynolds, J.H., Pasternak, T., Desimone, R. (2000) Attention increases sensitivity of V4 neurons. Neuron, 26, 703-714.
22. Spratling MW, Johnson MH. (2004) A feedback model of visual attention. J Cogn Neurosci. 16:219-237.
23. Treue, S., Maunsell, J.H. (1999) Effects of attention on the processing of motion in macaque middle temporal and medial superior temporal visual cortical areas, J. Neurosci., 19, 7591-7602.
24. Treue, S., Martínez Trujillo, J.C. (1999) Feature-based attention influences motion processing gain in macaque visual cortex. Nature, 399, 575-579.
25. Tsotsos JK, Culhane SM, Wai W, Lai Y, Davis N, Nuflo F (1995) Modeling visual attention via selective tuning. Artificial Intelligence, 78:507-545.
26. Usher, M., Niebur, E. (1996) Modeling the temporal dynamics of IT neurons in visual search: A mechanism for top-down selective attention. J. Cog. Neurosci., 8, 311-327.

Towards a Biologically Plausible Active Visual Search Model

Andrei Zaharescu, Albert L. Rothenstein, and John K. Tsotsos

Dept. of Computer Science and Centre for Vision Research,
York University, Toronto, Canada
{andreiz, albertlr, tsotsos}@cs.yorku.ca

Abstract. This paper proposes a neuronal-based solution to active visual search, that is, visual search for a given target in displays that are too large in spatial extent to be inspected covertly. Recent experimental data from behaving, fixating monkeys is used as a guide and this is the first model to incorporate such data. The strategy presented here includes novel components such as a representation of saccade history and of peripheral targets that is computed in an entirely separate stream from foveal attention. Although this presentation describes the prototype of this model and much work remains, preliminary results obtained from its implementation seem consistent with the behaviour exhibited in humans and macaque monkeys.

1 Motivation

Have you ever tried to imagine how would life be without moving our eyes? Most of the tasks that we perform in our everyday routine are greatly facilitated by our capacity to perform the fast eye movements known as saccades. Most simple actions, like finding the exit door, or locating the right button in the elevator, require a search of the visual environment and can be categorized as active visual search tasks, a search for given visual target that requires more than one eye fixation [1]. Several models ([2], [3], [4], [5]) [6, 7] have addressed aspects of attentive eye movements or active visual search with some success. At the present time, we are unaware of the existence of a computational biologically plausible model that compares favourably with human or primate active visual search performance. This contribution attempts to make significant inroads towards this goal. Although a general solution will be sketched, the demonstration is only for a limited case; the implementation of the full solution is in progress.

2 Background

Eye movements have been studied both psychophysically and neurophysiologically [8]. Several attempts have been made ([9], [10]) to formalize the performance characteristics of active search and to relate them to stimulus properties

L. Paletta et al. (Eds.): WAPCV 2004, LNCS 3368, pp. 133–147, 2005.

and cortical physiology. An outstanding accomplishment is the work of Motter et al. ([11][11], [12], [13]). They addressed the problem of active search for a target in a display of simple stimuli. The performance of monkeys on this task is documented and characterized. These characterizations form part of the basis for the model presented here. A second goal is to explore the use of the Selective Tuning Model (STM) [14] framework for this task, basing the work on the solution for one type of active fixation that was introduced in that paper.

2.1   Motters Experiments

Imagine the following experiment, shown in Fig. 1 below: the task is to locate a red bar oriented on a 135° angle (assume the trigonometric direction of the angle counterclockwise) from among a set of distracters oriented at 45° or 135°, coloured red or green. This experiment represents a conjunction search problem, where distracters and the target share common features [15]. The marked cross represents the initial point of fixation. The white rectangle denotes the portion of the overall scene that is visible within a single fixation. The remainder of the scene is depicted by the black rectangle that is not within the visual field of the subject. It can only be inspected by changes in fixation. Motter et al. ([11], [12], [13]) sought to discover the algorithm used by primates (macaque monkeys) to perform this task. As a result, they were able to characterize the solution used by macaques when performing active visual search tasks, the main components being summarized below: The probability of target detection falls off as a function of target eccentricity from fixation. The area within which targets are discovered is a function of the density of items around the target itself. In order for objects to be identified, some minimal distance must separate their representations in visual cortex. Therefore, because cortical magnification falls off as a function of eccentricity, the density of items around the target determines how close to the point of fixation the target must be before it can be identified. Color can be used to 'label' item relevancy. By knowing the probability of target detection as a function of eccentricity and assuming a random walk through relevant stimuli, they were able to account completely for search rate performance. No previous model of visual attention can account for this performance. The main goal of the research reported here is the attempt to cast such search characteristics into STM.

2.2   Selective Tuning Model

The Selective Tuning Model (STM) is a proposal for the explanation at the computational and behavioural levels of visual attention in humans and primates. Key characteristics of the model include:

1. A top-down coarse-to-fine Winner-Take-All selection process;
2. A unique Winner-Take-All (WTA) formulation with provable convergence properties;
3. A Winner-Take-All that is based on region rather than point selection;
4. A task-relevant inhibitory bias mechanism;

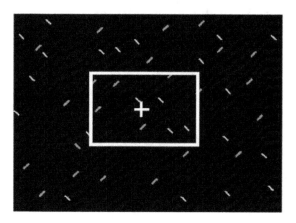

Fig. 1. Experimental setup

5. Selective inhibition in spatial and feature dimensions reducing signal interference leading to suppressive surrounds for attended items;
6. A task-specific executive controller.

The processing steps of the executive controller relevant for visual search tasks
follow:

1. Acquire target as appropriate for the task, store in working memory;
2. Apply top-down biases, inhibiting units that compute task irrelevant quantities;
3. See the stimulus, activating feature pyramids in a feed-forward manner;
4. Activate top-down WTA process at top layers of feature pyramids;
5. Implement a layer-by-layer top-down search through the hierarchical WTA
   based on the winners in the top layer;
6. After completion, permit time for refined stimulus computation to complete
   a second feedforward pass. Note that this feedforward refinement does not
   begin with the completion of the lowermost WTA process; rather, it occurs simultaneously with completing WTA processes (step 5) as they proceed downwards in the hierarchy. On completion of the lowermost WTA,
   some additional time is required for the completion of the feedforward
   refinement.
7. Extract output of top layers and place in working memory for task verifica-
   tion;
8. Inhibit pass zone connections to permit next most salient item to be processed;
9. Cycle through steps 4 - 8 as many times as required to satisfy the task.

Greater detail on STM and its mechanisms may be found in ([14], [16], [17]).

3   The Active Visual Search Model

We have developed a general-purpose active visual search model that tries to incorporate most of the observations noted by Motter et al. within the STM framework. It will be presented next, beginning with a general overview followed by a detailed description. The basic aspects of the problem to be solved include: it is a search problem; it involves eye movements; the probability of target detection is directly related to the target eccentricity; saccadic eye movement is guided by item density; the cortical magnification factor needs to be taken into account; target characteristics can bias the search process; for the Motter task, saccadic eye movements are guided by colour, but not orientation.

3.1   Overview

Active visual search denotes a target location task that involves eye movements. In STM, feature pyramids are constructed through pyramidal abstraction and, as argued in [14], this abstraction leads to natural separation between a central region where stimuli are veridically analyzed and a peripheral one where they are not. The obvious solution to this is to include an active strategy for determining whether to attend to a stimulus in this central region or to a peripheral one, and if the latter, to initiate a fixation shift in order to bring that peripheral item into the veridical central region. Thus, STM incorporates two main search sub-processes: a central search and a peripheral search. It is noted that this addresses only one aspect of attentive fixation shifts. Each search sub-process is guided by a corresponding instance of the attention mechanism. Covert attention allows inspection of stimuli in the foveal region, trying to locate the target, while the overt counterpart guides saccadic eye movements towards the next fixation once the covert search strategy fails to locate the target. It does not mean that covert attention does not have access to the peripheral information, but rather that there is an independent overt mechanism that utilizes the peripheral information only. Is there a neural correlate to this separation? There is strong biological evidence for the existence of an area in the parieto-occipital (PO) of macaque that receives retinotopical input from V1, V2, V3, V4 and MT and that is sensitive to the regions outside the central 10-degrees of visual angle ([18]). The receptive fields are large an order of magnitude larger than the ones in V2. It is a hypothesis of this solution that area PO may have the function of detecting peripheral targets in this context. Due to the boundary problem resulting from limited size receptive fields and the resulting non-veridical analysis in the periphery [14], area PO must have access to features that have not been processed through many layers of the hierarchy (they would not be valid otherwise). Thus, PO receives input only from early features in this model.

The covert search part of the system has been discussed in detail by in previous work [14] and has been briefly mentioned in the background section (the executive controller). The novelty is introduced by the addition of a saccadic eye movement control system. From this point onward we will refer to it as the Overt Control System. Also, we will call a (Feature) Plane a population of neurons with similar function and behaviour. The Overt Control System interacts with the rest of the system in the following ways:

- It receives an Eye Movement Bias from the Task Controller;
- It receives input from both the Foveal Target Plane and the Peripheral Target Plane, which operate in a STM Winner-Take-All network;
- It outputs a request to the Eye Movement Controller for a saccade.

A system block diagram is shown in Fig. 2, placing the new component, the Overt Control System, in the context of the full model and its implementation. The components relevant to the active visual search task are elaborated in the remainder of the paper.

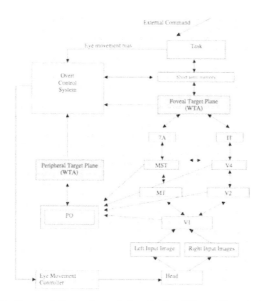

Fig. 2. General Overview of the Active Visual Search Model

3.2   The Overt Control System

In the current section we will describe the Overt Control System at a very general level, in terms of the interactions that take place among its components. Later we will provide a detailed description of the system tailored to address the limited task that is exemplified by Fig. 1. The Task Controller via the Eye Movement Bias triggers the Overt Control System, which is responsible for the choice of the most salient next saccade from the peripheral target plane. In doing so, it is important that it keeps track of previously inspected positions in the scene, since a return inspection would not yield to new information. This may be addressed by an inhibition of return mechanism (IOR; see [19]). IOR has been a staple of attention mechanisms since Koch & Ullman [20]. There, it was applied to a saliency map in order to inhibit a winner in a winner-take-all

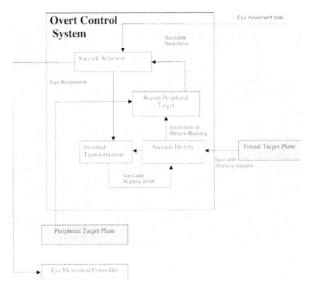

Fig. 3. Components of the Overt Control System

scheme, thus enabling the next strongest stimulus to emerge. Here, it must be in part a memory of what has been inspected and then used to inhibit a set of locations in order to avoid wasted fixations or oscillatory behaviour. The exact mechanism of IOR for saccadic eye movements is not currently known [19]. In the current model, we store the inhibition of return information in an entity named Saccade History. The Saccade History is constructed using a pool of neurons that encode position in the world; neurons will be on if they represent locations that have been previously visited for the task, and off otherwise. There are all reset to off at the end of the task. The saccade selection task from the peripheral target plane will be a basic winner-take all process selecting a winning location with all features competing equally. The feature locations are inhibitively biased by the locations in the Saccade History and indirectly by the concentration of distracters, due to RF size at the PO level. The Task Controller manages the overall behaviour (saccade selection) of the Overt Control System via an Eye Movement Bias, which influences equally in a multiplicative fashion all the participating neurons in the next saccade Winner Take All process. Thus, a high value of the bias would facilitate a new saccade, whereas a low value would prevent a new saccade from taking place. In case of a saccade winner, the command will be sent to the Head controller. Upon the selection of a winner, a new saccade will be issued. The Saccade History information is updated so that it integrates the current fixation point and also so that it maintains the information in a world system of coordinates that will be consistent with the new eye shift. It means that for each eye shift, there will be an associated saccade history shift. These basic components are depicted in Fig. 3. Each of the Overt Control System components will be discussed in greater detail in the next section, using the example task proposed in the opening section.

4 Simple Conjunction Search

This section will describe how the above framework can account for the active visual search performance observed by Motter and colleagues for a simple conjunction search as described in the Background. All the details of the framework will not be included here; only the elements directly involved in the solution of such tasks (Fig. 1) will be shown. Several other experimental results are currently being tested within the framework. The section begins with a brief description of the basic neural simulator that forms the substrate for the model implementation. The active visual search model has been implemented using TarzaNN, a General Purpose Neuronal Network Simulator [21] as a foundation. In TarzaNN, feature computations are organized in feature planes, interconnected by filters that define the receptive field properties of the neurons in the planes.

4.1 Eye Movement Controller

The current simulation implements a virtual eye movement controller, which only presents a part of the virtual 2D world to the system. In our case, referring to the setup of Fig. 1, the virtual world would represent the entire image, and the current input the area within the rectangle. An eye movement would correspond to a shift of the rectangle in the desired direction. This will eventually be replaced by a controller to a robotic stereo camera system.

4.2 The Input

The Foveal and Peripheral Target Planes are computed through a hierarchy of filters as shown in Fig. 2. The input image is first processed in order to incorporate the cortical magnification factor, the effect of large receptive fields and decreasing concentration of neural processing. There are mathematical descriptions to fit the biological data [22] that result in an increasing blur applied to the input image. In our model we have used the following two parameter complex log cortical magnification function (CMF):

$$CMF(z) = k\frac{log(z+a)}{log(z+b)}, a \cong 0.3, b \cong 50 \qquad (1)$$

where z, a and b represent degrees of visual angle, and k a normalization factor (the maximum in our case). In the very center of the fovea there is no need to blur the original pixels, but as eccentricity (z) increases, the level of blurring will increase.

4.3 Feature Target Planes

In the implementation of the current example, we focused mainly on testing the validity of the Overt Control System, and not to implement a full-featured STM model, including all the areas described in the opening diagram (i.e. V1, V2, V4, IT, MT, MST, 7a; however see [14], [17] for several of these). We have modeled only a skeleton of feature planes sufficient to provide the Overt Control System

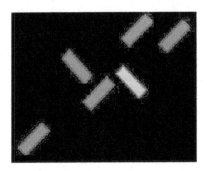
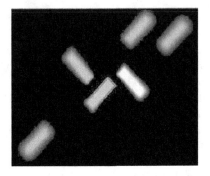

Fig. 4. Input Image - the fixated subset of a typical full scene

Fig. 5. Transformed input image (with the equivalent cortical magnification operator)

with the required information, namely the Foveal and Peripheral Target Planes, together with the eye movement bias. A port of the full featured STM model within TarzaNN is in progress. In order to focus this description on the Motter task, the only features discussed here will be colour and orientation. Hence, a sensible feature plane organization is the following:

- The foveal region hierarchy will have colour selective feature planes (R=r (g+b)/2, G=g-(r+b)/2, B=b-(r+g)/2), and for each colour, orientation selective feature planes (for 45° and 135°).
- The peripheral region will contain only colour selective feature planes.

In order to simulate the density distribution, which would occur normally at the top layers of a STM pyramid, a 2D Gaussian filter is applied. Please note that the current feature plane layout is tailored specifically for the above-mentioned experiment and it represents a very stripped-down version of realistic model (i.e. the peripheral region only receives the red-coloured filtered input image in reality, colour is coded in both fovea and para-fovea). Figures 4 through 10 show a feature plane walk-through starting with the sample input image. The captions provide the processing stage description in each case.

4.4   Saccade History

Earlier we introduced the idea of a Saccade History, to incorporate the task memory in the system. We will associate a feature plane with the concept. In the current simulation we are only considering a 2D version of the world, thus restricting the 3D space to a plane. However, at least at the hypothetical level, a generalization of the concept is immediately available for 3 dimensions.

The most efficient implementation of such a saccade history would be to only remember a list of the previously visited locations. However, it is not clear what a neural correlate of a list might be. Perhaps a more straightforward solution would be to model it as a scaled down version of the external world. This represents another hypothesis of the model, that a neural correlate of such a representation

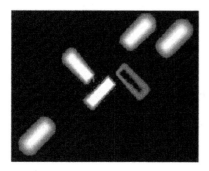

Fig. 6. R=r-(g+b)/2 colour filter plane

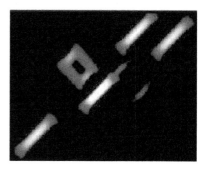

Fig. 7. 45° oriented region plane with R=r-(g+b)/2 plane as input

Fig. 8. 135° oriented region plane with R=r-(g+b)/2 plane as input

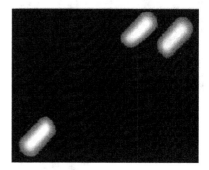

Fig. 9. Peripheral Region plane from R=r-(g+b)/2 colour filter plane

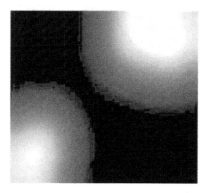

Fig. 10. Salient peripheral locations based on input as in figure 9 computing a stimulus density distribution using a Gaussian filter

exists with this functionality. The Saccade History encodes regions visited overtly by x/y location.

Assume the following Saccade History of previously executed saccades as shown in Figure 11. By convention, the centre of the map is at current eye fixation. A contour showing the currently fixation falls is depicted by a rectangle in Fig. 12. If a previous saccade location overlaps with a target, it is inhibited. For the initial Peripheral Target Plane, the Biased Peripheral Target will appear as shown in Fig. 13. The inhibition produced by the Saccade History plane is never total. Sometimes, distracters could be visited more than once in accomplishing a task [13].

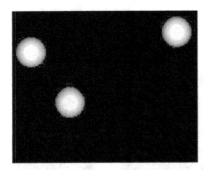   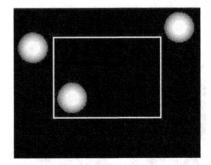

Fig. 11. Assumed Past Saccade History   **Fig. 12.** Assumed Past Saccade History

Fig. 13. Inhibition of return Biased Peripheral Target

4.5   Saccade History Feature Updating

The Saccade History Plane is updated to include the current fixation for each shift, as shown at the top of Fig. 14 . In this particular implementation of the system we have chosen to encode the location of fixation only. The properties of the attended stimulus (provided by the Foveal Feature Plane) are not taken into account. Whether location and feature qualities are both represented is unclear; there is little neurophysiological guidance on this currently. Further studies and experiments are needed in order provide an answer.

4.6   Saccade History Transformation

The center of the Saccade History Plane always encodes the current fixation. When a saccade is performed, the point of fixation moves in the world. In order to properly represent all previously fixated regions and maintain the world coordinates, a corresponding shift has to take place of all represented fixation locations (this may be related to the attentional shifts reported in [23] but it is too early at this stage to know). If we consider the saccade as a transformation given by the parameters (Tx,Ty), where these represent Euclidean distances in pixels for this example, then we need to perform the inverse of that transformation (Tx, Ty) to the Saccade History Plane. This is due to the fact that the world and the observer have systems of reference of opposite polarities. If the initial Saccade History Plane is filtered via a pool of neurons with receptive field properties described by a 2D Gaussian spread function shifted in the opposite direction of the saccade, we obtain the desired result. The winning saccade triggers the shifted corresponding inverted set of neurons, while inhibiting the others, thus ensuring that the appropriate transformation takes place. Biological evidence suggests that such a correspondence between the saccade target and a world transformation exists [24]. As more factors are taken into account, such as the 3rd dimension, head movements, and world movements, additional transformations will be needed to maintain the saccade history, using essentially the same mechanism. Pouget et al. [25] provide a detailed biological implementation of coordinate transformations that we are considering including in a future version of the system.

4.7   Saccade Selection

The most salient feature in the biased peripheral feature plane encodes the coordinate of the next saccade. This will be established via a Winner Take All process operating on the biased peripheral features selecting the strongest responding one.

As the Saccade History Plane is shifted, some information is discarded, since data is shifted outside the boundaries of the space, as illustrated in Fig. 14. The coordinates of the next saccade are sent to the Eye Movement Controller.

5   Results

The overall system generated very satisfying results, in accordance with the goals. Fig. 15 presents the result of an average trial (in terms of number of fixations). We present the results of the same search task when the there is no colour discrimination and the inhibition of return mechanism is turned off. As it can be observed in Fig.16, the inhibition or return mechanism is crucial for the task, so that the process will not get into an infinite loop. We show another run where the next saccade fixation point is chosen randomly. As it can be seen in Fig. 17, it took over 200 trials to obtain the same result. The performance of the system is dictated by some of the system control variables. They are listed below,

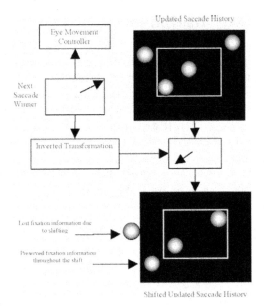

Fig. 14. Saccade History Shifting. This example illustrates how, upon a Saccade History shift, certain previously visited locations information can get lost

together with the relationship in which they influence the overall performance of the system:

- The size of the window into the virtual world (directly proportional);
- The accuracy of the next fixation in terms of selecting the region with the most distracters (inversely related to the size and spread of the 2D Gaussian filter as illustrated in Fig.10);
- The size of the inhibited region in the Saccade History Plane (directly proportional);
- The amount of inhibition in the Saccade History Plane (directly proportional).
- The ratio between dimensions of The Saccade History Plane and the Input Image

The probability of target detection is indeed directly related to the target eccentricity, and that follows as a direct consequence of the fact that cortical magnification is taken into account, and the solution uses neuronal response thresholding for detection. The current system represents our first approach to the problem, and at the present time we cannot provide a detailed analysis of its performance with regards to real test data. Relevant statistical data needs to be generated with a large set of example cases in order to be able to plot it against the graphs provided by Motter et al. for direct comparison.

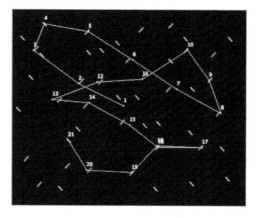

Fig. 15. Sample Run

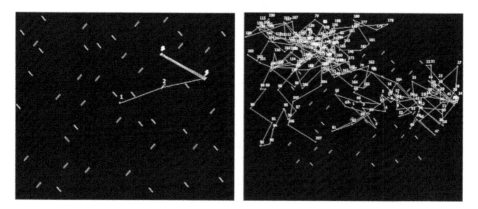

Fig. 16. Example without Colour Discrimination and Inhibition of Return

Fig. 17. Sample Run using a random saccade fixation

6   Conclusions

The system represents our first approach to solve the active visual search problem in a biologically plausible fashion. It integrates findings about the specifics of the task and about biological constraints. The proposed solution to active visual search contains several novel components. The newly added Overt Control System is capable of guiding the saccadic eye movements taking into account peripheral distracter characteristics and by incorporating inhibition of return information. The general system can solve a whole class of active visual search problems as long as the appropriate Peripheral Target and a Foveal Target Plane are provided (see Fig.2). The Overt Control System represents and important milestone in the evolution of the Selective Tuning Model, exposing it to a whole new set of problems and challenges From the biological standpoint, the current implementation of the system exhibits a certain weakness in the assumption of the 2D structure of the Saccade History Plane. We are not claiming that there exists a one-to-one correspondence between the model proposed and a biological hierarchy of neuronal constructs. However, from a functional standpoint, a similar mechanism needs to encode the previously visited locations of the world and store them in some type of memory (short term, or more specialized location based memory/world map). Further research will need to address this outstanding issue.

Acknowledgements

This research was funded by the Institute for Robotics and Intelligent Systems, a Government of Canada Network of Centres of Excellence and by Communication and Information Technology Ontario, a Province of Ontario Centre of Excellence. JKT holds the Canada Research Chair in Computational Vision.

References

1. Tsotsos, J.K.: On the relative complexity of passive vs active visual search. International Journal of Computer Vision **7** (1992) 127 – 141
2. Wolfe, J.: Guided search 2.0. a revised model of visual search. Psychonomic Bulletin & Review **1** (1994) 202–238
3. Itti, L., Koch, C.: A saliency-based search mechanism for overt and covert shifts of visual attention. Vision Research **40** (2000) 1489–1506
4. Rao, R.P., Zelinsky, G.J., Hayhoe, M.M., Ballarad, D.H.: Eye movements in iconic visual search. Vision Research **42** (2002) 1447–1463
5. Hamker, F.M.: The reentry hypothesis: linking eye movements to visual perception. Journal of Vision **3** (2003) 808–816
6. Lanyon, L.J., Denham, S.L.: A model of active visual search with object-based attention guiding scan paths. Neural Networks **in press** (2004)
7. Lanyon, L.J., Denham, S.L.: A biased competition computational model of spatial and object-based attention mediating active visual search. Neurocomputing **in press** (2004)
8. Carpenter, R.: 8, Vision and Visual Dysfunction. In: Eye Movements. Volume 8. J. Cronly-Dillon (ed.), MacMillan Press (1991) 95–137
9. Triesman, A., Sato, S.: Conjunction search revisited. J. Experimental Psychology: Human perception and Performance **16** (1990) 459–478
10. Lennie, P., Trevarthen, C., van Essen, D., Wassle, H.: Parallel processing of visual information. San Diego: Academic Press (1990)
11. Motter, B., Belky, E.J.: The zone of focal attention during active visual search. Vision Research **38** (1998) 1007–1022
12. Motter, B.C., Belky, E.J.: The guidance of eye movements during active visual search. Vision Research **38** (1998) 1805–1815
13. Motter, B.C., Holsapple, J.W.: Cortical image density determines the probability of target discovery during active search. Vision Research **40** (2000) 1131–1322
14. Tsotsos, J.K., Culhane, S., Wai, W., Lai, Y., Davis, N., Nuflo, F.: Modeling visual attention via selective tuning. Artifical Intelligence **78** (1995) 507–547

15. Treisman, A., Gelade, G.: A feature integration theory of attention. Cognition Psychology **12** (1980) 97–136
16. Tsotsos, J.K.: Analyzing vision at the complexity level. Behavioural Sciences **13** (1990) 423–445
17. Tsotsos, J.K., Liu, Y., Martinez-Trujillo, J., Pomplun, M., Simine, E., Zhou, K.: Attending to visual motion. submitted (2004)
18. Galletti, C., Gamberini, M., Kutz, D.F., Fattori, P., Luppino, G., Matelli, M.: The cortical connections of area v6: an occipito-parietal network processing visual information. European Journal of Neuroscience **13** (2001) 1572–1588
19. Klein, R.M.: Inhibition of return. Trends in Cognitive Sciences **4** (2000) 138–147
20. Koch, C., Ullman, S.: Shifts in the selective visual attention: Towards the underlying neural circuitry. Human Neurobiology **5** (1985) 219–227
21. Rothenstein, A.L., Zaharescu, A., Tsotsos, J.K.: TarzaNN: A general purpose neural network simulator for visual attention modeling. In: Workshop on Attention and performance in computational vision. (2004)
22. Schwartz., E.L.: Topographic mapping in primate visual cortex: Anatomical and computational approaches. Visual Science and Engineering: Models and Applications **43** (1994)
23. Duhamel, J.R., Colby, C.L., Goldberg, M.E.: The updating of the representation of visual space in parietal cortex by intended eye-movements. Science **255** (1992) 90–92
24. Fukushima, K., Yamanobe, T., Shinmei, Y., Fukushima, J.: Predictive responses of periarcuate pursuit neurons to visual target motion. Exp. Brain Res (2002) 104–120.
25. Pouget, A., Sejnowski, T.J.: Spatial transformations in the parietal cortex using basis functions. Journal of Cognitive Neuroscience **9** (1997) 222–237

Modeling Grouping Through Interactions Between Top-Down and Bottom-Up Processes: The Grouping and Selective Attention for Identification Model (G-SAIM)

Dietmar Heinke, Yaoru Sun, and Glyn W. Humphreys

School of Psychology, University of Birmingham,
Birmingham B15, 2TT, United Kingdom

Abstract. We preset a new approach to modelling grouping in a highly-parallel and flexible system. The system is is based on the Selective Attention for Identification model (SAIM) [1], but extends it by incorporating feature extraction and grouping processes: the Grouping and Selective for Identification for Identification model (G-SAIM). The main grouping mechanism is implemented in a layered grouping-selection network. In this network activation spreads across similar adjacent pixels in a bottom-up manner based on similarity-modulated excitatory connections. This spread of activation is controlled by top-down connections from stored knowledge. These top-down connections assign different groups within a known object to different layers of the grouping-selection network in a way that the spatial relationship between the groups is maintained. In addition the top-down connections allow multiple stances of the same objects to be selected from an image. In contrast, selection operates on single objects when the multiple stimuli present are different. This implementation of grouping within and between objects matches a broad range of experimental data on human visual attention. Moreover, as G-SAIM maintains crucial features of SAIM, earlier modeling successes are expected to be repeated.

1   Introduction

SAIM (Selective Attention for Identification Model) is a connectionist model of human visual attention[1]. SAIM's behaviour is controlled by interactions between processing units within and between modules that compete to control access to stored representations for translation invariant object recognition to take place. SAIM gives a qualitative account of a range of psychological phenomena on both normal and disordered attention. Simulations on normal attention are consistent with psychological data on: two-object costs on selection, effects of object familiarity on selection, global precedence, spatial cueing both within and between objects, and inhibition of return. When SAIM was lesioned by distorting weights within the selection module, it also demonstrated both unilateral

L. Paletta et al. (Eds.): WAPCV 2004, LNCS 3368, pp. 148–158, 2005.

neglect and spatial extinction [2], depending on the type and extent of the lesion. Different lesions also produced view-centred and object-centred neglect, capturing the finding that both forms of neglect can occur not only in different patients but also within a single patient with multiple lesions. In essence, SAIM suggested that attentional effects on human behaviour result from competitive interactions in visual selection for object recognition, whilst neurological disorders of selection are due to imbalanced competition following damage to areas of the brain modulating access to stored knowledge. In [3] we compared SAIM with the most important models on human selective attention (e.g. MORSEL [4], SERR [5], saliency-based models e.g. [6] and biased-competition models, e.g. [7]) and showed that SAIM covers widest range of experimental evidence.

Interestingly, this comparison also highlighted the fact that few models of human visual selection incorporate grouping processes, despite the fact that grouping plays an important role in the processing of visual information in humans (see [8] for a review). SERR [5] implemented a simple grouping process only based on the identify of objects, not taking into account other forms of grouping, e.g. similarity-based grouping. In its original form SAIM too employed very simplistic grouping processes. In particular, units in the selection network supported activity in neighbouring units that could contain a proximal element, leading to grouping by proximity. Here, we present the results from new simulations showing that SAIM's architecture can be extended to incorporate more sophisticated forms of grouping sensitive to the similarity of the feature present in the display (bottom-up). In addition, experimental evidence shows that grouping is also influenced by top-down factors (e.g. [9]). Therefore, grouping-SAIM does not have a hard-wired coding of specific conjunctions of elements, but rather uses a flexible grouping and selection procedure which operates in interaction between image-based similarity constraints and top-down knowledge imposed by an object recognition system. Moreover, this form of grouping does not only operate within objects but it can also occur across multiple objects, linking separate objects into representations encompassing the whole display. This matches data on human search, where people can respond at the level of the whole display to multiple, homogenous stimuli (e.g. [10]). Interestingly, the approach to grouping and selection in the revised model produce a form of size-invariant object representation as an emergent property.

2   SAIM

2.1   Overview

Figure 1 gives an overview of G-SAIM's architecture and highlights the modular structure of the model. In the first stage features are extracted from the input image. The contents network maps a section of grouped features into a smaller Focus of Attention (FOA), a process modulated by the grouping-selection network. In addition the mapping of the contents network into the FOA is translation-invariant, enabling G-SAIM to perform translation-invariant object recognition. The grouping-selection network has a multilayered structure

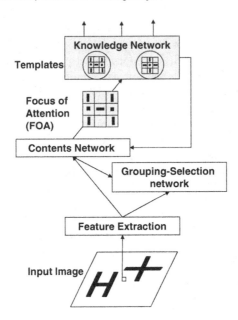

Fig. 1. Architecture of G-SAIM. The depiction of the contents of FOA and the templates of the knowledge network illustrates feature values arranged in a spatial grid. Each grid element in the FOA represents the average feature values of a group, as computed by the contents network

where each layer corresponds to a particular location in the FOA. The operation of the grouping-selection network is controlled by competitive and cooperative interactions. These interactions ensure that adjacent locations with similar features are grouped together, whilst adjacent locations with dissimilar features are separated into different groups. Grouped items are represented by conjoint activity within a layer of the grouping-selection network and different groups are represented in different layers. At the top end of the model, the knowledge network identifies the contents of the FOA using template matching. Importantly, in addition to these feedforward functions there are feedback connections between each module. The feedback connection from the knowledge network to the contents network aims at activating only known patterns in the FOA. The connection from the contents network into the grouping-selection network modulates the grouping and selection process so that patterns matching the content of the FOA are preferred over unknown objects.

The design of G-SAIM's network follows the idea of soft constraint satisfaction in neural networks based on "energy minimization" through gradient descent [11]:

$$\tau \dot{x}_i = -\frac{\partial E(y_i)}{\partial y_i} \tag{1}$$

with x_i being the internal activation of the unit i and y_i the output activation of the unit i. Both activations are linked through a non-linear function.

In G-SAIM the "energy minimization" approach is applied in the following way: Each module in SAIM performs a pre-defined task (e.g. the knowledge network has to identify the pattern in the FOA). In turn, each task describes allowed states of activations in the network after the network has converged. These states then define the minima in an energy function. To ensure that the model as a whole satisfies each constraint, set by each network, the energy functions of each network are added together to form a global energy function for the whole system. The minima in the energy function is found via gradient descent, as proposed by [11], leading to a recurrent network structure between the modules. At this point it should be noted that the gradient descent is only one of many possible algorithms that could be applied to find the minimal energy. In our earlier work this approach turned out to be sufficient for modeling psychological data. For technical applications of G-SAIM alternative approaches might need to be considered. In the following sections the energy functions for each network are stated. The global energy function and the gradient descent mechanism are omitted, since they are clearly defined by the subcomponents of the energy function.

2.2   Feature Extraction

The feature extraction results in a three-dimensional feature vector: horizontal and vertical lines and the image itself. The lines are detected by filtering the image with 3x3 filters ($\begin{smallmatrix} -2 & +1 & -2 \\ -2 & +1 & -2 \\ -2 & +1 & -2 \end{smallmatrix}$ for vertical lines and its transposed version for horizontal lines). The feature vector is noted as f_{ij}^n hereafter, with indices i and j refereing to image locations and n to the feature dimension. This feature extraction process provides an approximation of simple cell responses in V1. As becomes obvious in the following sections, the use of this simple feature extraction mechanism is not of theoretical value in its own right and arises primilary from practical consideration (e.g., the duration of any simulations). In principle, a more biologically realistic feature extraction process can be substituted (e.g. using Gabor filter).

2.3   Contents Network

The energy function for the contents network is:

$$E^{CN}(\mathbf{y}^{\mathbf{GN}}, \mathbf{y}^{\mathbf{CN}}) = \sum_{ijlm} (y_{lmn}^{CN} - f_{ij}^n)^2 \cdot y_{lmij}^{GN} \tag{2}$$

y_{lmij}^{GN} is the activation of units in the grouping-selection network and y_{lmn}^{CN} is the activation of units in the contents network. Here and in all the following equations the indices l and m refer to locations in the FOA. The term $(y_{lmn}^{CN} - f_{ij}^n)^2$ ensures that the units in the contents network match the feature values in the input image. The term y_{lmij}^{GN} ensures that the contents of the FOA only reflect the average feature values of a region selected by the grouping-selection network ($y_{lmij}^{GN} = 1$). Additionally, since setting an arbitrary choice of y_{lmij}^{GN}s to 1 allows any location to be routed from the feature level to the FOA level, the contents network enables a translation-invariant mapping. It should be noted that the energy function of the contents network results in a feedback connection to the grouping-selection network. This connection provides the grouping-selection network with featural information and its relative positions within the FOA.

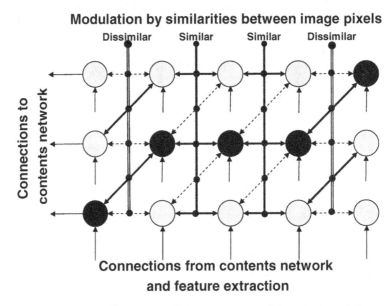

Fig. 2. One-dimensional illustration of the structure and functioning of the grouping-selection network (see text for details)

2.4   Grouping-Selection Network

The mapping from the retina to the FOA is mediated by the grouping-selection network. In order to achieve successful grouping and selection, the grouping-selection network has to fulfill certain constraints when it modulates the mapping process. These constraints are that: (i) the content of a image location should be mapped only once into the FOA; (ii) units whose related image pixels are similar appear in the same layer (implementing similarity grouping) (iii) dissimilar features are routed into separate layers; (iv) neighbouring groups should appear adjacent in the FOA. Figure 2 illustrates the functioning of the grouping-selection network. For the sake of clarity only excitatory connections are depicted. The grouping-selection network has a layered structure where each each layer is connected to one node in the contents network. Within each layer and between each layer there are excitatory connections depicted as dotted and bolt lines. These connections are modulated by the similarity between adjacent pixels. The more similar two pixels are the higher is the excitation of the within-layer connection (bolt lines in Fig. 2) and the lower the excitation of the between-layer connections, vice versa.

In the Figure 2 the first two pixels and the last two pixels are dissimilar from each other, whereas the pixels in the middle are similar to each other. Consequently, the first and the last excitatory connection is strong between layers (diagonal) whereas the excitation within the layer is strong in the middle pixels. During the process of energy minimization process in G-SAIM this particular connectivity pattern can lead to an activation of units on the diagonal for the first and the last unit and an activation of units within one layer in the middle part (black circles). Hence, the pixels in the middle are grouped together, whereas the dissimilar pixel are separated from the middle pixels by activating different layers.

The similarity between pixels is determined in the following way:

$$s_{ijsr} = h(\sum_n (f_{ij}^n - f_{i+s,j+r}^n)^2) \tag{3}$$

$$h(x) = \frac{1}{1 + e^{(a \cdot x + b)}}$$

The parameter for the nonlinearity $g(x)$ is set so that the similarity values range from 0 to 1 (0 = dissimilar; 1 = similar) for given a range feature values.

The energy function for activating similar pixels within a layer of the grouping-selection network is:

$$E^{exc\_sim}(\mathbf{y^{GN}}) = -\sum_{lmij} y_{lmij}^{GN} \sum_{\substack{s=-M \\ s \neq 0}}^{M} \sum_{\substack{r=-M \\ M \neq 0}}^{1} g_{rs} \cdot s_{ijrs} \cdot y_{l,m,i+r,j+s}^{GN} \tag{4}$$

To prevent a region from spreading across the whole image an "inhibitory" energy function was introduced in each layer of the grouping-selection network:

$$E^{inh\_sim}(\mathbf{y^{GN}}) = \sum_{lm}(\sum_{ij} y_{lmij}^{GN})^2 \tag{5}$$

The coefficients g_{rs} drop off in a Gaussian shape to order to reduce the influence of pixels that are further apart. In essence the two equations above represent an implementation of a similarity-modulated Amari-network [12]. Amari showed that under certain conditions gaussian-shaped excitatory connections within a layer lead to contiguous areas of activation. In case of the grouping-selection network the shape of these areas are influenced by similarity in order to implement grouping.

To ensure that dissimilar adjacent pixels are assigned to separate layers, the following energy function was introduced:

$$E^{exc\_dis} = \sum_{lmij} y_{lmij}^{GN} \sum_{\substack{s=-1 \\ s \neq 0}}^{1} \sum_{\substack{r=-1 \\ r \neq 0}}^{1} (1 - s_{ijrs}) \cdot y_{l+r,m+s,i+r,j+s}^{GN} \tag{6}$$

The term $(1 - s_{ijrs})$ is a measure for the dissimilarity between pixels.

The following term prevents units at the same image location to be activated in different layers (constraint (i)):

$$E^{inh}(\mathbf{y^{GN}}) = \sum_{ij}(\sum_{lm} y_{lmij}^{GN})^2 \qquad (7)$$

The Constraint (iv) was implemented by connecting all adjacent units which are not connected by similarity modulated connections in an inhibitory way. The energy functions above all include a stable state at zero activation. To prevent G-SAIM from converging into this state an "offset"-term was added to the network:

$$E^{offset}(\mathbf{y^{GN}}) = -\sum_{ijlm}(y_{lmij}^{GN}) \qquad (8)$$

At this point the general nature of the four constraints implemented by the grouping-selection network should be noted. The constraints are not specific about which feature belongs to which layer. The exact assignment of the groups to the different layers in the network is achieved by the top-down influence from the contents network and the knowledge network. These networks feed featural information from known objects back into the grouping-selection network, setting the featural preference of each layer for the groups. The top-down influence, paired with the generality of the grouping constraint, is at the heart of the flexible grouping approach in G-SAIM.

2.5   Knowledge Network

The energy function of the knowledge network is defined as

$$E^{KN}(\mathbf{y^{KN}}, \mathbf{y^{CN}}) = a^{KN}(\sum_k y^{KN} - 1)^2 -$$

$$\sum_k \begin{cases} \sum_{lmn} y_k^{KN} \cdot y_{lmn}^{KN} & \text{If } \mathbf{w^k\text{'}dont select'} \\ \sum_{lmn} w_{lm}^* \cdot (y_{lmn}^{CN} - w_{lmn}^k)^2 \cdot y_k^{KN} & \text{otherwise} \end{cases} \qquad (9)$$

The index k refers to the template unit. The term $(\sum_k y^{KN} - 1)^2$ restricts the knowledge network to activate only one template unit [13]. Each grid field of a template has a flag attached to it indicating if this particular grid field contains information relevant for the object. For instance, in case of the cross the areas between the arms do not belong to the objects; consequently these positions are assigned a "don't select" flag. In future simulations the "don't select"-flag will served to deal with cluttered scenes. In case of an activated "don't select"-flag the term $\sum_{lmn} y_k^{KN} \cdot y_{lmn}^{KN}$ aims at suppressing any activation in the contents network and grouping-selection network. On the contrary, the term $\sum_{lmn}(y_{lmn}^{CN} - w_{lmn}^k)^2 \cdot y_k^{KN}$ ensures that the best-matching template unit is activated. a^{KN} and b^{KN} weight these constraints against each other.

3   Results and Discussion

Two sets of simulations were run to test the grouping mechanism in the grouping-selection network. Two objects were used, a cross and an H. The knowledge network had these two objects as templates (see Fig. 1 for an illustration). In the first set of simulations input images with one object (cross or H) were used (see Fig. 3). The results show that G-SAIM successfully segments parts of the objects (H and cross) into sub-groups, as indicated by the fact that the arms of the H/cross appear in separated layers of the grouping-selection network. This is a direct outcome of the feature extraction of lines which leads to similar feature values along the arms of the H and dissimilar feature values at its cross points.

Grouping-selection network:

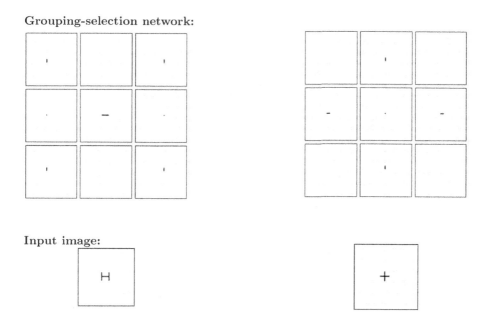

Input image:

Fig. 3. Simulation results with a single object in the input image. As a result of the grouping mechanism implemented in the grouping-selection network, each layer represents a different group of the H.cross, indicating a grouping within an object

These simulations illustrate the fact that the layers in the grouping-selection network are not connected firmly to a certain feature. For instance, in the simulation with cross as input the centre layer represents the cross point whereas for the H the centre layer encodes the horizontal bar of the H. Compared to other neural network approaches to grouping (e.g. [14]), where groups in one layer for each feature, the grouping process in G-SAIM is both efficient and flexible. Moreover, it allows the model to maintain vital information about the spatial relation between groups within objects in a natural way. For instance, for the H spatial relations between its arms are important information to distinguish it from other objects,

Grouping-selection network:

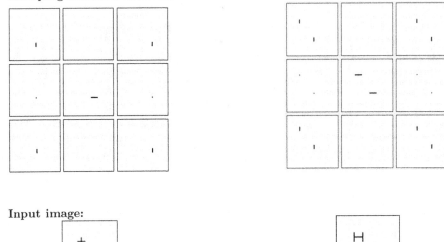

Input image:

Fig. 4. Simulation results. The result on the left illustrates a case with two different objects (H and cross). The cross is suppressed and the H is selected. The simulation result on the right shows a result of grouping the same object as the two Hs are selected together

e.g. the cross. In G-SAIM this information is simply encoded by assigning the different segments to different layers of the grouping-selection network. It is unclear how such this form of coding could be achieved in a grouping mechanism where is a layer per feature. For instance in the case of the H the arms might be represented in one layer, but then an additional mechanism would be necessary to encode the spatial relations between the parts.

In the second set of simulations images with two objects were used. In these simulations the knowledge network had a slight bias towards activating the template unit for the H. Figure 4 show the simulation results of the grouping-selection network for two input images: on the left a cross together with an H and on the right two Hs. With two different objects, the cross and the H, G-SAIM selects the H and suppresses the cross. Hence, as in old SAIM, G-SAIM still performs object selection. The second simulation examines G-SAIM's behaviour with two Hs in the input image (Fig. 4 right). In this case G-SAIM selected both objects together. The decision concerning whether two objects are selected together or not is influenced through the constraints implemented in the contents network. If units in the grouping-selection network correspond to the same features in the input image, the activation of these units contribute to the minimization of the energy function. In contrast, if the units correspond to different feature values, the activation in the grouping-selection network is suppressed, as was the case for the cross-H simulation. These two contrasting behaviours, selection of identical objects and competition of different objects, match extensive evidence from human search (e.g. [10]) and from phenomena such as visual extinction in patients with parietal lesions (e.g. [15]).

Finally, it should be noted that the representation of objects in the FOA possesses interesting properties. In the FOA the average feature values for each group are represented. Since the features within the groups are similarly determined by the grouping-selection network, averaging provides a good approximation of the features values within the group. Consequently, when the size of the regions changes and the features in the regions stay the same, the representation in the FOA does not change, leading to a size-invariant representation of objects. Moreover, the size-invariant representation in the FOA is very compact, e.g. it represents the cross through a small number of units.

4   Conclusion

G-SAIM represents the first step towards modeling grouping and selective attention in an integrative approach. The core mechanism for the grouping process is activation spreading, modulated by bottom-up similarity between pixels and top-down influence from object knowledge. G-SAIM simulates the grouping of regions within objects as well as grouping across multiple objects. To the best our knowledge no other models integrate these two forms of behaviour. For instance, the saliency-based approach to selective attention [6] does not have a grouping mechanism. The same is true for the biased-competition approaches (e.g. [7]). In terms of architecture G-SAIM is closest to the dynamic shifter circuits [16], but again this model does not contain a feature extraction and grouping mechanisms. Moreover, other models on grouping utilize similar activation spreading mechanisms (e.g. [14]. However, it is not clear how these models could cope with multiple object situations.

In addition it should be noted that G-SAIM also keeps crucial features of the old SAIM, especially the interaction of competitive and cooperative processes and the multilayer structure of the grouping-selection network. Both elements were responsible for the modeling successes of old SAIM. Hence, we expect G-SAIM to be able reproduce the simulation results of old SAIM, whilst having added capability in simulating grouping effects.

In future work we will aim at replacing the present feature extraction process by a more biologically-plausible approach (e.g. using a Gabor filter) and at simulating psychological data on grouping and attention, especially experimental evidence on the interactions between grouping and attention [9].

Acknowledgment

This work was supported by grants from the EPSRC (UK) to the authors.

References

1. D. Heinke and G. W. Humphreys. Attention, spatial representation and visual neglect: Simulating emergent attention and spatial memory in the Selective Attention for Identification Model (SAIM). *Psychological Review*, 110(1):29–87, 2003.
2. K. M. Heilman and E. Valenstein. Mechanisms underlying hemispatial neglect. *Annals of Neurology*, 5:166–170, 1979.
3. D. Heinke and G. W. Humphreys. Computational Models of Visual Selective Attention: A Review. In G. Houghton, editor, *Connectionist Models in Psychology*. Psychology Press, in press.
4. M. C. Mozer and M. Sitton. Computational modeling of spatial attention. In H. Pashler, editor, *Attention*, pages 341–393. London:Psychology Press, 1998.
5. G. W. Humphreys and H. J. Müller. SEarch via Recursive Rejection (SERR): A Connectionist Model of Visual Search. *Cognitive Psychology*, 25:43–110, 1993.
6. J. M. Wolfe. Guided Search 2.0 A revised model of visual search. *Psychonomic Bulletin & Review*, 1(2):202–238, 1994.
7. G. Deco and J. Zihl. Top-down selective visual attention: A neurodynamical approach. *Visual Cognition*, 8(1):119–140, 2001.
8. S. E. Palmer. *Vision Science – Photons to Phenomenology*. The MIT Press, 1999.
9. J. Driver, G. Davis, C. Russell, M. Turatto, and E. Freeman. Segmentation, attention and phenomenal visual objects. *Cognition*, 80:61–95, 2001.
10. J. Duncan and G. W. Humphreys. Visual Search and Stimulus Similarity. *Psychological Review*, 96(3):433–458, 1989.
11. J. J. Hopfield and D.W. Tank. "Neural" Computation of Decisions in Optimazation Problems. *Biological Cybernetics*, 52:141–152, 1985.
12. Shun-ichi Amari. Dynamics of Pattern Formation in Lateral-Inhibition Type Neural Fields. *Biological Cybernetics*, 27:77–87, 1977.
13. E. Mjolsness and C. Garrett. Algebraic Transformations of Objective Functions. *Neural Networks*, 3:651–669, 1990.
14. S. Grossberg, E. Mingolla, and W. D. Ross. A Neural Theory of Attentive Visual Search: Interactions of Boundary, Surface, Spatial and Object Representation. *Psychological Review*, 101:470–489, 1994.
15. I. Gilchrist, G. W. Humphreys, and M. J. Riddoch. Grouping and Extinction: Evidence for Low-Level Modulation of Selection. *Cognitive Neuropsychology*, 13:1223–1256, 1996.
16. B. Olshausen, C. H. Anderson, and D. C. Van Essen. A Multiscale Dynamic Routing Circuit for Forming Size-and Position- Invariant Object Representations. *Journal of Computational Neuroscience*, 2:45–62, 1995.

TarzaNN : A General Purpose Neural Network Simulator for Visual Attention Modeling

Albert L. Rothenstein[1], Andrei Zaharescu[1], and John K. Tsotsos[1]

Dept. of Computer Science and Centre for Vision Research,
York University, Toronto, Canada
{andreiz, albertlr, tsotsos}@cs.yorku.ca

Abstract. A number of computational models of visual attention exist, but making comparisons is difficult due to the incompatible implementations and levels at which the simulations are conducted. To address this issue, we have developed a general-purpose neural network simulator that allows all of these models to be implemented in a unified framework. The simulator allows for the distributed execution of models, in a heterogeneous environment. Graphical tools are provided for the development of models by non-programmers and a common model description format facilitates the exchange of models. In this paper we will present the design of the simulator and results that demonstrate its generality.

1 Introduction

Even though attention is a pervasive phenomenon in primate vision, surprisingly little agreement exists on its definition, role and mechanisms, due at least in part to the wide variety of investigative methods. As elsewhere in neuroscience, computational modeling has an important role to play by being the only technique that can bridge the gap between these methods [1] and provide answers to questions that are beyond the reach of current direct investigative methods.

A number of computational models of primate visual attention have appeared over the past two decades (see [2] for a review). While all models share several fundamental assumptions, each is based on a unique hypothesis and method. Each seems to provide a satisfactory explanation for several experimental observations. However, a detailed comparative analysis of the existing models, that is, a comparison with each of the models subjected to the same input data set in order to both verify the published performance and to push the models to their limits, has never been undertaken. Such an analysis would be invaluable: comparative, computational testing procedures would be established for the first time, successful modeling ideas would be confirmed, weaknesses identified, and new directions for development discovered. The goal would be to validate the models by testing them against existing knowledge of the primate attentional system; the experimental stimuli and task definitions which led to that knowledge would form the basis for the development of the test data sets.

L. Paletta et al. (Eds.): WAPCV 2004, LNCS 3368, pp. 159–167, 2005.

In order to facilitate this analysis and to provide the research community with a common software platform, we have developed a general purpose, extensible, neural network simulator geared towards the computational modeling of visual attention. The simulator allows for the distributed execution of models in a heterogeneous environment. Its associated tools allow non-programmers to develop and test computational models, and a common model description format facilitates the exchange of models between research groups. The simulation results can be presented in a variety of formats, from activation maps to the equivalent of single-unit recordings and fMRI.

This paper starts with a discussion of the design of the simulator, from the perspective of the requirements imposed by existing computational models of visual attention. This is followed by a review of some of the hardware and performance issues that were taken into account. The paper continues with a description of the simulator user interface and the modeling tools that accompany it and concludes with a discussion of the system and the current state of the project, including preliminary results, items considered for future development, and a comparison with other neural network simulators.

2   Simulator Requirements and Design

The wide variety of computational models of visual attention creates unique challenges for a unified framework. For example, computer science inspired models use simple, generic McCulloch-Pitts [3] type neurons (e.g. [4], [5]), sometimes in combination with specialized neurons (e.g. gating neurons in [4]), in large-scale simulations; others use neurons modeled by realistic differential equations in minimalist, proof of concept networks (e.g. [6]) or large-scale systems with spike density functions (e.g. [7]) or spiking (e.g. [8]).

Object-oriented techniques have been applied to the specification of the various categories of neurons in the system in a very flexible and easily extendable framework. Currently, the system provides simple McCulloch-Pitts type neurons and two categories modeled by differential equations: Wilson-Cowan for spike density functions and Hodgkin-Huxley for spiking neurons [9]. New types of neurons can be added by simply subclassing the existing, high-level models and specializing the behaviour as desired. In particular, new kinds of neurons modeled by differential equations can be added by creating a class that describes their equations.

All large-scale computational models of visual attention share the notion of a field of neurons with identical properties and connectivity. For the rest of the paper we will refer to these fields as feature planes. Thus, a model consists of a number of feature planes, interconnected by filters that define the receptive field properties of the neurons in the plane. This division of the models into feature planes and filters allows for very efficient implementation of the core of the simulator as a convolution engine.

Different models and levels of simulation require distinct methods of communication and synchronization between feature planes. Three such methods have been implemented: lock step, asynchronous and synchronous. The lock step method corresponds to traditional neural network behaviour, where the various components perform computational steps according to an external clock determined by the slowest component. The asynchronous method allows each computation to be performed at its own pace, without any coordination except, of course, for the locking needed to ensure that data transferred between feature planes is in a consistent state. This method is the closest we can come to a biological, decentralized system, and it is the most appropriate for the more realistic simulations that use neurons defined by differential equations. If there are significant speed differences between feature planes, this method means that the fastest one will perform several steps on the same data. To handle this scenario, we have introduced the synchronous communication method, which is identical to the previous one, except for the fact that feature planes notify their dependents when new results are available and the dependents will not perform computations unless notified.

Due to the wide variety of computing platforms available, portability was a major concern in the design and implementation of the simulator. While portability alone would seem to suggest Java as the programming language of choice, performance considerations make it impractical. The code is written in ANSI C++, and the only external requirement is the highly portable Qt package from Trolltech. The simulator was developed and tested on Mac OS X, Windows, Linux and Solaris, and in a heterogeneous cluster composed of computers running various versions of these operating systems.

Given the stated goal of creating a system to be used by many research groups, it was natural to adopt an open source development process. The whole project is available on the Internet (http://www.tarzaNN.org), and researchers can participate in the development process either directly, by submitting code through CVS, which is used for source control (http://www.cvshome.org/), or by requesting features and reporting bugs through Bugzilla (http://www.bugzilla. org/) and its web interface. Documentation, also available on the project web site, is generated automatically using doxygen (www.doxygen.org/).

3   Performance Considerations

Performance is key to the success of any simulator, and in this case performance has two aspects. First and most obvious, model execution has to be as close as possible to real time, especially in the computationally very expensive case of differential equation models. A second, and by no means less important consideration is the speed and ease with which models can be developed and modified (see Section 4).

The structuring of the models in feature planes connected through filters made convolutions the main computation in the system, and this allowed us to apply classical techniques from image processing to speed up the calculations. Here we will mention only two of these techniques. Filters are analyzed using linear algebra and, if possible, separated using singular value decomposition (SVD), which transforms the two dimensional convolution into a sequence of two one dimensional convolutions. In cases where decomposition is not possible, the designer has the option of defining the filter as a linear composition of simpler filters that are applied in parallel, increasing the chances that the simpler filters are candidates for the SVD algorithm. An example of this is the difference-of-Gaussians filter, which is not decomposable, but is a linear combination of decomposable filters. The system also exploits the vector arithmetic units built into PowerPC chips. All complex object-oriented software systems have to address the performance problems related to the use of virtual functions and the fact that they impose late binding (i.e. the code to be executed has to be determined at run time). Due to the fact that we require particular flexibility in the combination of neuron types and synchronization strategies, this issue has been addressed early in the design of our simulator. To alleviate this problem, policy-based design relying on C++ templates [10] was used extensively, allowing us to assemble complex behaviour in the feature plane classes out of many simple, specialized classes without the overhead of the classical object oriented approach.

One important observation is that any visual attention model comprises of a number of clusters of high connectivity with relatively sparse connectivity between clusters. Generally, the clusters correspond to the visual areas in the primate brain with dense local interconnections in the form of inhibitory pools, winner-take-all and center-surround circuits. This structure makes it possible to distribute the computation across a group of computers. Feature planes are represented on remote machines by proxy objects that handle the communication in its entirety, making the whole process completely transparent.

4   Tooling and Interfaces

Two of the key requirements for the simulator are accessibility to non-programmers and support for collaborative research, by allowing easy exchange of models between groups. To facilitate this, models are described by using a common description format based on the XML language. A tool is being developed to graphically describe the models, allowing researchers who are not familiar with programming to use the system. XML files are automatically generated. The graphical designer has a look and feel that is very similar to that of existing tools for drawing diagrams, with typical drag-and-drop toolbars for the definition of feature planes and their interconnections. Double-clicking on the feature planes opens the properties dialog, where users can specify their size, type of neurons they contain and their parameters, and other characteristics. Similarly, in the link properties dialog users can determine the size, type, and parameters of filters.

The simulator user interface presents the results of the computations in three formats. The default view presents each feature plane as an image, with shades of gray corresponding to local activations. The main window of the application is presented in Fig. 1, with the input image and a number of activation maps

Fig. 1. The main window, presenting activation maps for the feature planes used in simulating the results presented by Itti et al. [5]. In their model, feature map activity is combined at each location, giving rise to a topographic saliency map. A winner-take-all network detects the most salient location. The top-left corner is the input image and the bottom-right the saliency map. The other sub-windows represent (left to right, top to bottom) activation maps for center-surround feature planes for: the image luminance, the red, the green and the blue colour channels, vertical, 45 degrees left, horizontal, and 45 degrees right orientations. The saliency map is a linear combination of the center-surround feature maps

being displayed. The time-course view presents the temporal evolution of the output of individual neurons within a feature plane. Depending on the nature of the model neuron, these can correspond to spike density functions or action potentials (for Wilson-Cowan or Hodgkin-Huxley neurons, respectively [9]). In Fig. 2-left, we present two spike density functions, corresponding to two neurons in a competitive network. The neuron represented in red corresponds to the winning neuron, with a characteristic response profile, while the green trace corresponds to a neuron that is inhibited by the proximity of the winner. Finally, the fMRI view presents a comparison between the current state of one or more feature planes and a snapshot taken during a different run. This comparison can be performed off-line, between two saved snapshots of activity, or in real time, between a saved snapshot and the currently executing network. The fMRI view integrating activations across all feature planes corresponding to a brain area should closely match observations made in human subjects. Fig. 2-right presents the static comparison between two activations (left hand side of

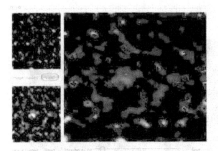

Fig. 2. Left: Spike density functions for two Wilson-Cowan neurons [9] involved in a competitive interaction (mutual inhibition through an interneuron, not represented here). The red trace corresponds to the winning neuron, and shows the characteristic response of a neuron, with the high stimulus onset response, followed by the longer adaptation period. The green trace shows the response of a neuron that was inhibited as a result of the competition. Right: Static functional MRI view corresponding to the difference between the two activation maps on the left. The bottom activation map is the response of non-Fourier V4 concentric units [1] to a Glass pattern (see Figure 4), while the top is the response of the same feature map to a random dot pattern. The central red area in the fMRI view represents the higher activation corresponding to the detected concentric pattern, while the other red and blue areas represent stronger responses to the Glass pattern and the random pattern, respectively. These weaker secondary areas can be eliminated by adjusting the threshold at the bottom of the figure

the image), corresponding to the response of model non-Fourier (second order) V4 concentric units [1] to a Glass pattern [11] vs. a random dot pattern, with the characteristic colour scheme. A threshold slider is available at the bottom of the window, this can be used to eliminate statistically insignificant image components. Note that the feature planes communicate with the various views through an observer/notification scheme, so it is very easy to add custom views if a specific application needs them.

For an example of a large-scale system that was implemented in TarzaNN, see [12].

The simulator accepts input images in a wide variety of file formats, or directly from a camera, and includes mechanisms to control the motion of the camera, if applicable.

5   Discussion and Conclusions

A number of computational models of primate visual attention have appeared over the past two decades, but due to the different initial assumptions and requirements, each modeling effort has started from scratch, and the designs reflect

Fig. 3. Left   input image, Right   saliency map. This model reproduces the results presented by Itti et al. (Figure 3 in [5]). The model identifies the phone pole as the most salient area of the image, followed by the traffic sign. See Fig. 1 for details and a view of the whole network

these requirements and the particular software and hardware present in that particular lab.

In this paper we presented the design principles behind a general purpose neural network simulator specifically aimed at computational modeling of visual attention and we discussed hardware and performance issues and how they influenced design and implementation. We also presented the user interface, with the many ways in which simulation results can be presented to the user. The tools that make the simulator accessible to non-programmers and allow for collaborative research, by facilitating the exchange of models between groups, were also introduced.

The flexibility of the simulator and the XML-based model description have allowed us to very quickly obtain implementations of the three major compu-

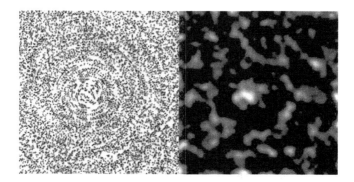

Fig. 4. Left   input Glass pattern, Right   activation map of V4 concentric unit feature plane. This network implements the non-Fourier (second order) processing presented by Wilson (Figure 6 in [1]). As can be observed in the activation map, the highest activation corresponds to the V4 concentric units corresponding to the center of the illusory circles. See also Figure 2, for a comparison between the networks response to the Glass pattern vs. a random dot pattern

tational models of visual attention ([7], [4], [5]). Fig. 3 (details extracted from Fig. 1) corresponds to part of Fig. 3 in [5]. Fig. 3-left represents the input image, while Fig. 3-right is the activation of the saliency map feature plane. The largest of these models is a subset of the Selective Tuning Model for motion [13], which consists of 210 feature planes organized in 4 areas (the full model will include 620 feature planes for the feedforward network, plus at least that number for the winner-take all circuits). Note that the applicability of the simulator is not limited to modeling visual attention, see for example Fig. 4, where we have reproduced the ability of model non-Fourier (second order) V4 concentric units proposed in [1] to detect Glass patterns [11]. In terms of performance, at the current level of optimization, on a dual processor 2.0Ghz PowerMac G5, compiled using gcc version 3.3, a sub-network composed of 3 feature planes of Wilson-Cowan neurons described by three differential equations each, connected through 5 filters, performs one step in 40ms - two 128 by 128 and one 20 by 20 feature planes, or 33,168 neurons, which means roughly 800,000 neuron updates per second.

Distributed execution of models is currently the focus of significant work, and we should soon have the ability to automatically distribute feature planes across a network, task that currently has to be done manually. In terms of the model design tool, future plans include the ability to automatically generate sets of feature planes that differ in certain dimensions (e.g. orientation selective planes for a given set of angles and spatial resolutions). Also, the tool should allow designers to group feature planes into areas and specify how these should be distributed across the network of computers currently this has to be done manually.

TarzaNN distinguishes itself from the other neural network simulators available by uniquely bringing together a series of characteristics. Portability is extremely important for collaborative research, and many simulators are limited to either Windows or one or more Unix platforms (e.g. SNNS [14], Genesis [15], iNVT [16], etc). Also, many simulators have built-in limitations that make them inappropriate for the simulation of some visual attention models, by limiting neuron types (e.g. Amygdala [17], PDP++ [18]) or network configurations (e.g. iNVT [16], NeuralWorks [19]). General-purpose neural network simulators (e.g. Neural Network Toolbox for Matlab [20]) require programming skills for anything but the most common architectures, and performance is limited by the lack of facilities for distributed computations. Usually there is a trade-off between flexibility and the ability to design large-scale networks.

The main original contributions of the work presented in this paper are the novel object oriented framework for defining neuron properties, the configurable methods of communication and synchronization between layers and the standard-based method of describing networks. The neuron framework allows users to define model neurons at any level of complexity desired, and even test the behaviour of the same network with different models. The synchronization and communication infrastructure makes it possible to simulate both traditional computer science neural networks and biologically plausible models. The XML based network description is an important contribution in at least two ways: it allows for easy interchange of models between researchers and it makes the automatic generation of regular structures possible without a need for programming skills.

References

1. Wilson, H.: Non-fourier cortical processes in texture, form and motion perception. In: Cerebral Cortex. 13. Ulinski et al. edn. Kluwer Academic/Plenum Publishers (1999)
2. Itti, L., Koch, C.: Computational modeling of visual attention. Nature Reviews Neuroscience **2** (2001) 194–203
3. McCulloch, W.S., Pitts, W.: A logical calculus of ideas immanent in nervous activity. Bulletin of Mathematical Biophysics **5** (1943) 115–133
4. Tsotsos, J.K., Culhane, S., Wai, W., Lai, Y., Davis, N., Nuflo, F.: Modeling visual attention via selective tuning. Artifical Intelligence **78** (1995) 507–547
5. Itti, L., Koch, C., Nierbur, E.: A model of saliency-based visual attention for rapid scene analysis. IEEE Transactions on Pattern Analysis and Machine Intelligence **20** (1998) 1254–1259
6. Reynolds, J.H., Chelazzi, L., Desimone, R.: Competitive mechanisms subserve attention in macaque areas v2 and v4. The Journal of Neuroscience **19** (1999) 1736–1753
7. Rolls, E.T., Deco, G.: Computational Neuroscience of Vision. Oxford Univ. Press (2002)
8. Lee, K.W., Buxton, H., Feng, J.F.: Selective attention for cue-guided search using a spiking neural network. In Paletta, L., Humphreys, G.W., Fisher, R.B., eds.: Proc. International Workshop on Attention and Performance in Computer Vision. (2003) 55–63
9. Wilson, H.R.: Spikes, decisions and actions: dynamical foundations of neuroscience. Oxford University Press, Oxford UK (1999)
10. Alexandrescu, A.: Modern C++ Design. Addison-Wesley (2001)
11. Glass, L.: Moire effect from random dots. Nature **223** (1969) 578–580
12. Zaharescu, A., Rothenstein, A.L., Tsotsos, J.K.: Towards a biologically plausible active visual search model. In et al. (Eds.), L.P., ed.: WACPV, Prague (2004) 67–74
13. Tsotsos, J.K., Liu, Y., Martinez-Trujillo, J., Pomplun, M., Simine, E., Zhou, K.: Attending to motion: Localizing and classifying motion patterns in image sequence. In Bülthoff, H.H., Lee, S.W., Poggio, T., Wallraven, C., eds.: Proceedings of the Second International Workshop on Biologically Motivated Computer Vision, Tübingen, Germany (2004) 439–452
14. URL: (Stuttgart neural network simulator http://www-ra.informatik.uni-tuebingen.de/snns/)
15. URL: (Genesis http://www.genesis-sim.org/genesis/)
16. URL: (ilab neuromorphic vision c++ toolkit http://ilab.usc.edu/toolkit/)
17. URL: (Amygdala http://amygdala.sourceforge.net/)
18. PDP++: (http://www.cnbc.cmu.edu/resources/pdp++)
19. URL: (Neuralworks http://www.neuralware.com/)
20. URL: (Neural network toolbox for matlab http://www.mathworks.com/products/neuralnet/)

Visual Attention for Object Recognition in Spatial 3D Data

Simone Frintrop, Andreas Nüchter, and Hartmut Surmann

Fraunhofer Institut für Autonome Intelligente Systeme,
Schloss Birlinghoven, 53754 Sankt Augustin, Germany
simone.frintrop@ais.fraunhofer.de
http://www.ais.fraunhofer.de/~frintrop/attention.html

Abstract. In this paper, we present a new recognition system for the fast detection and classification of objects in spatial 3D data. The system consists of two main components: A biologically motivated attention system and a fast classifier. Input is provided by a 3D laser scanner, mounted on an autonomous mobile robot, that acquires illumination independent range and reflectance data. These are rendered into images and fed into the attention system that detects regions of potential interest. The classifier is applied only to a region of interest, yielding a significantly faster classification that requires only 30% of the time of an exhaustive search. Furthermore, both the attention and the classification system benefit from the fusion of the bi-modal data, considering more object properties for the detection of regions of interest and a lower false detection rate in classification.

1 Introduction

Object recognition and classification belong to the hardest problems in computer vision and have been intensely researched [1]. Generally, for a given domain a large number of different object classes has to be considered. Although fast classifiers have been built recently [2], it is time consuming to apply many classifiers to an image. To preserve high quality of recognition despite of limited time, the input set has to be reduced. One approach is to confine classification to image regions of potential interest found by attentional mechanisms.

In human vision, attention helps identify relevant parts of a scene and focus processing on corresponding sensory input. Psychological work shows evidence that different features, like color, orientations, and motion, are determined in parallel, coding the saliency of different regions [3]. Many computational models of attention are inspired by these findings [4, 5, 6].

In this paper, we present a new system for the fast detection and recognition of objects in spatial 3D data, using attentional mechanisms as a front end for object recognition (Fig. 1). Input is provided by a 3D laser scanner, mounted on an autonomous mobile robot. The scanner yields range as well as reflectance data in a single 3D scan pass [7]. Both data modalities are transformed into images

L. Paletta et al. (Eds.): WAPCV 2004, LNCS 3368, pp. 168–182, 2005.

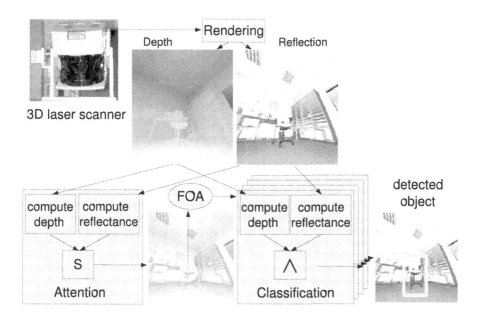

Fig. 1. The recognition system. Two laser modes are provided by a 3D laser scanner, rendered into images and fed into an attention system which computes a focus of attention (FOA) from the data of both modes. The classifier searches for objects only in the FOA-region in depth and reflectance image and combines the results by an appropriate connection. The rectangles in the result image (right) depict a detected object

and fed into a visual attention system based on one of the standard models of visual attention by Koch & Ullman [8]. In both laser images, the system detects regions which are salient according to intensity and orientations. Finally, the focus of attention is sequentially directed to the most salient regions.

A focus region is searched for objects by a cascade of classifiers built originally for face detection by Viola and Jones [2]. Each classifier is composed of several simple classifiers containing edge, line or center surround features. The classifiers are applied to both laser modes. We show how the classification is significantly sped up by concentrating on regions of interest with a time saving that increases proportionally with the number of object classes. The performance of the system is investigated on the example of finding chairs in an office environment. The future goal will be a flexible vision system that is able to find different objects in order of their saliency. The recognized objects will be registered in semantic 3D maps, automatically created by the mobile robot.

The fusion of two sensor modalities is performed in analogy to humans who use information from all senses. Different qualities of the modes enable to utilize their respective advantages, e.g., there is a high probability that discontinuities in range data correspond to object boundaries what facilitates the detection of objects: an object producing the same intensity like its background is difficult to detect in an intensity image, but easily in the range data. Additionally, misclassification of shadows, mirrored objects and wall paintings is avoided. On the other hand, a flat object, e.g., a sign on a wall, is likely not to be detected in the range but in the reflectance image. The respective qualities of the modes significantly improve the performance of both systems by considering more object properties for focus computation and a lower rate of false detections in classification. Furthermore, the scanner modalities are illumination independent, i.e., they are the same in sunshine as in complete darkness and no reflection artifacts confuse the recognition.

The presented architecture introduces an new approach for object recognition, however, parts of it have already been investigated. Pessoa and Exel combine attention and classification, but whereas we detect salient objects in complex scenes, they focus attention on discriminative parts of pre-segmented objects [9]. Miau, Papageorgiou and Itti detect pedestrians on attentionally focused image regions using a support vector machine algorithm [10]; however, their approach is computationally much more expensive and lacks real-time abilities. Object recognition in range data has been considered by Johnson and Hebert using an ICP algorithm for registration of 3D shapes [11], an approach extended in [12]; in contrast to our method, both use local, memory consuming surface signatures based on prior created mesh representations of the objects.

The paper is organized as follows: Section 2 describes the 3D laser scanner. In section 3 we introduce the attentional system and in 4 the object classification. Section 5 presents the experiments performed by the combination of attention and classification and discusses the results. Finally, section 6 concludes the paper.

2   The Multi-modal 3D Laser Scanner

The data acquisition in our experiments was performed with a 3D laser range finder (top of Fig. 1, [7]). It is built on the basis of a 2D range finder by extension with a mount and a small servomotor. The scanner works according to the time-of-flight principle: It sends out a laser beam and measures the returning reflected light. This yields two kinds of data: The distance of the scanned object (range data) and the intensity of the reflected light (reflectance data).

One horizontal slice is scanned by serially sending out laser beams using a rotating mirror. A 3D scan is performed by step-rotating the 2D scanner around a horizontal axis scanning one horizontal slice after the other. The area of $180°(\text{h}) \times 120°(\text{v})$ is scanned with different horizontal (181, 361, 721 pts) and vertical (210, 420 pts) resolutions. To visualize the 3D data, a viewer program based on OpenGL has been implemented. The program projects a 3D scene to the image plane, such that the data can be drawn and inspected from every perspective. Typical images have a size of 300×300 pixels. The depth information of the 3D data is visualized as a gray-scale image: small depth values are represented as bright intensities and large depth values as dark ones.

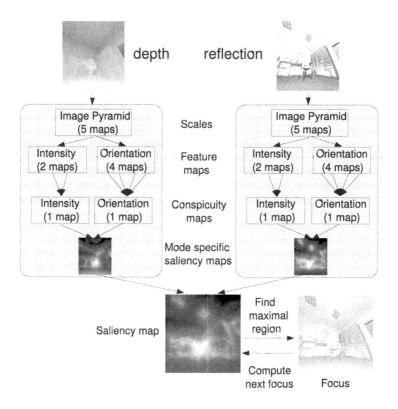

Fig. 2. The Bimodal Laser-Based Attention System (BILAS). Depth and reflectance images rendered from the laser data are processed independently. Conspicuities according to intensity and orientation are determined and fused into a mode-specific saliency map. After combining both of these maps, a focus of attention (FOA) is directed to the most salient region. Inhibition of return (IOR) enables the computation of the next FOA

3   The Laser-Based Attention System

The Bimodal Laser-Based Attention System (BILAS) detects salient regions in laser data by simulating eye movements. Inspired by the psychological work of Treisman et al. [3], we determine conspicuities of different features, intensity and orientation, in a bottom-up, data-driven manner. These conspicuities are fused into a saliency map and the focus of attention is directed to the brightest, most salient point in this map. Finally, the region surrounding this point is inhibited, allowing the computation of the next FOA.

The attention system is shown in Fig. 2 (cf. [13]); it is built on principles of one of the standard models of visual attention by Koch & Ullman [8] that is used by many computational attention systems [4, 6, 14, 15]. The implementation of the system is influenced by the Neuromorphic Vision Toolkit (NVT) by Itti et al. [4] that is publicly available and can be used for comparative experiments (cf. [13]). BILAS contains several major differences as compared to the NVT. In the following, we will describe our system in detail emphasizing the differences between both approaches.

The main difference to existing models is the capability of BILAS to process data of different sensor modalities simultaneously, an ability not available in any other attention system the authors know about. In humans, eye movements are not only influenced by vision but also by other senses and the fusion of different cues competing for attention is an essential part of human attention. The sensor modalities used in this work are depth and reflectance values provided by the 3D laser scanner; in future work, we will use camera data additionally. The system computes saliencies for every mode in parallel and finally fuses them into a single saliency map.

Feature Computations

On images of both laser modalities, five different scales (0–4) are computed by Gaussian pyramids, which successively low-pass filter and subsample the input image; so scale $i+1$ has half the width and height of scale i. Feature computations on different scales enable the detection of salient regions with different sizes. In the NVT, 9 scales are used but the scales 5 to 8 are only used for implementation details (see below) so our approach yields the same performance with fewer scales. As features, we consider intensity and orientation.

The intensity feature maps are created by center-surround mechanisms which compute the intensity differences between image regions and their surroundings. These mechanisms simulate cells of the human visual system responding to intensity contrasts. The center c is given by a pixel in one of the scales $2 - 4$, the surround s is determined by computing the average of the surrounding pixels for two different sizes of surrounds with a radius of 3 resp. 7 pixels. According to the human system, we determine two kinds of center-surround differences: the on-center-off-surround difference $d_{(\text{on-off})}(c, s) = c - s$, responding strongly to bright regions on a dark background, and the off-center-on-surround difference $d_{(\text{off-on})}(c, s) = s - c$, responding strongly to dark regions on a bright background. This yields $2 \times 6 = 12$ intensity feature maps. The six maps for each center-surround variation are summed up by inter-scale addition, i.e., all maps are resized to scale 2 and then added up pixel by pixel. This yields 2 intensity maps.

The computations differ from those in the NVT, since we compute on-center-off-surround and off-center-on-surround differences separately. In the NVT, these computations are combined by taking the absolute value $|c - s|$. This approach is a faster approximation of the above solution but yields some problems. Firstly, a correct intensity pop-out is not warranted as is depicted in Fig. 3, top. The white object pops out in the computation with BILAS but not with the NVT. The reason is the amplification of maps with few peaks (see below). Secondly, if top-down influences are integrated into the system, a bias for dark-on-bright or bright-on-dark is not possible in the combined approach but in the separated

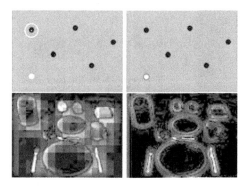

Fig. 3. Differences of NVT (left) and BILAS (right). Top: The white pop-out is not detected by the NVT (left) but by BILAS (right). Only separating the on-center-off-surround difference from the off-center-on-surround difference enables the pop-out. Bottom: Two intensity maps of a breakfast table scene, computed by the NVT (left) and by BILAS (right). The square-based structure in the left image resulting from taking the difference between two scales can be seen clearly, the right image shows a much more accurate solution

one. This is for instance an important aspect when the robots searches for an open door, visible as a dark region in the depth image. The two approaches vary also in the computation of the differences themselves. In the NVT, the differences are determined by subtracting two scales at a time, e.g. $I_6 = scale(4) - scale(8)$. Our approach results in a slightly slower computation but is much more accurate (cf. fig. 3, bottom) and needs fewer scales.

The orientation maps are obtained by creating four oriented Gabor pyramids detecting bar-like features of the orientations $\{0\,°, 45\,°, 90\,°, 135\,°\}$. In contrast to Itti et al., we do not use the center-surround technique for computing the orientation maps. The Gabor Filters already provide maps, showing strong responses in regions of the preferred orientation and weak ones elsewhere, which is exactly the information needed. Finally, the maps $2 - 4$ of each pyramid are summed up by inter-scale addition. This yields four orientation feature maps of scale 2, one for each orientation.

Fusing Saliencies

All feature maps of one feature are combined into a conspicuity map, using inter-scale addition for the intensity maps and pure pixel addition for the orientation maps. The intensity and the orientation conspicuity maps are summed up to a mode-specific saliency map, one representing depth and one reflection mode. These are finally summed up to the single saliency map S.

If the summation of maps was done in a straightforward manner, all maps would have the same influence. That means, that if there are many maps, the influence of each map is very small and its values do not contribute much to

Fig. 4. A trajectory of the first 6 foci of attention, generated by the attention system

the summed map. To prevent this effect, we have to determine the most important maps and give them a higher influence. To enable pop-out effects, i.e., immediate detection of regions that differ in one feature, important maps are those that have few popping-out salient regions. These maps are determined by counting the number of local maxima in a map that exceed a certain threshold. To weight maps according to the number of peaks, each map is divided by the square root of the number of local maxima m over a certain threshold: $w(\text{map}) = \text{map}/\sqrt{m}$.

The Focus of Attention

To determine the most salient location in S, the brightest point is located in a straightforward way instead of using a winner-take all network as proposed by Itti et al. While losing biological plausibility, the maximum is found even though with less computational resources. Starting from the most salient point, region growing finds recursively all neighbors with similar values within a certain threshold. The width and height of this region yield an elliptic FOA, considering size and shape of the salient region in contrast to the circular fixed-sized foci of most other systems.

Finally, an inhibition of return mechanism (IOR) is applied to the focused region by resetting the corresponding values in the saliency map, enabling the computation of the next FOA. Fig. 4 shows an example run of the system; to depict the output of the system, we present a trajectory for a single camera image instead of two laser images.

If two laser images are supplied as input, the attention system benefits from the depth as well as from the reflectance data, since these data modes complement each other: An object producing the same intensity like its background may not be detected in a gray-scale image, but in the range data. On the other hand, a flat object – e.g. a poster on a wall or a letter on a desk – is likely not to be detected in the depth but in the reflectance image (cf. [16]).

Fig. 5. Left: Edge, line, diagonal and center surround features used for classification. Right: The computation of the sum of pixels in the shaded region is based on four integral image lookups: $F(x,y,h,w) = I(x+w,y+h) - I(x,y+h) - I(x+w,y) + I(x,y)$. Feature values are calculated by subtractions of these values weighted with the areas of the black and white parts

4   Object Classification

Recently, Viola and Jones have proposed a boosted cascade of simple classifiers for fast face detection [2]. Inspired by these ideas, we detect office chairs in 3D range and reflectance data using a cascade of classifiers composed of several simple classifiers.

Feature Detection Using Integral Images

The six basic features used for classification are shown in Fig. 5 (left); they have the same structure as the Haar basis functions also considered in [17, 2, 18]. The base resolution of the object detector is 20×40 pixels, thus the set of possible features in this area is very large (361760 features). A single feature is effectively computed on input images using integral images [2], also known as summed area tables [18]. An integral image I is an intermediate representation for the image and contains the sum of gray-scale pixel values of an image N:

$$I(x,y) = \sum_{x'=0}^{x} \sum_{y'=0}^{y} N(x',y').$$

The integral image is computed recursively by the formula: $I(x,y) = I(x,y-1) + I(x-1,y) + N(x,y) - I(x-1,y-1)$ with $I(-1,y) = I(x,-1) = 0$, requiring only one scan over the input data. This representation allows the computation of a feature value using several integral image lookups and weighted subtractions (Fig. 5 right). To detect a feature, a threshold is required which is automatically determined during a fitting process, such that a minimum number of examples are misclassified.

Learning Classification Functions

The Gentle Ada Boost Algorithm is a variant of the powerful boosting learning technique [19]. It is used to select a set of simple features to achieve a given detection and error rate. The various Ada Boost algorithms differ in the update scheme of the weights. According to Lienhart et al., the Gentle Ada Boost Algorithm is currently the most successful learning procedure for face detection applications [18].

Learning is based on N weighted training examples $(x_i, y_i), i \in \{1 \ldots N\}$, where x_i are the images and $y_i \in \{-1, 1\}$ the supervised classified output. At the beginning, the weights w_i are initialized with $w_i = 1/N$. Three steps are repeated to select simple features until a given detection rate d is reached: First, every simple feature is fit to the data. Hereby, the error e is evaluated with respect to the weights w_i. Second, the best feature classifier h_t is chosen for the classification function and the counter t is increased. Finally, the weights are updated with $w_i := w_i \cdot e^{-y_i h_t(x_i)}$ and renormalized.

The final output of the classifier is $\text{sign}(\sum_{t=1}^{T} h_t(x))$, with $h(x) = \alpha$, if $x \geq$ thr. and $h(x) = \beta$ otherwise. α and β are the outputs of the fitted simple feature classifiers, that depend on the assigned weights, the expected error and the classifier size. Next, a cascade based on these classifiers is built.

The Cascade of Classifiers

The performance of one classifier is not suitable for object classification, since it produces a high hit rate, e.g., 0.999, and error rate, e.g., 0.5. Nevertheless, the hit rate is much higher than the error rate. To construct an overall good classifier, several classifiers are arranged in a cascade, i.e., a degenerated decision tree. In every stage of the cascade, a decision is made whether the image contains the object or not. This computation reduces both rates. Since the hit rate is close to one, their multiplication results also in a value close to one, while the multiplication of the smaller error rates approaches zero. Furthermore, the whole classification process speeds up, because the whole cascade is rarely needed. Fig. 6 shows an example cascade of classifiers for detecting chairs in depth images.

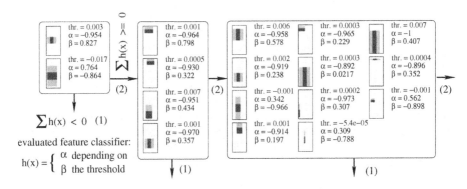

Fig. 6. The first three stages of a cascade of classifiers to detect an office chair in depth data. Every stage contains several simple classifiers that use Haar-like features

To learn an effective cascade, the classification function $h(x)$ is learned for every stage until the required hit rate is reached. The process continues with the next stage using only the currently misclassified examples. The number of features used in each classifier increases with additional stages (cf. Fig. 6).

The detection of an object is done by laying a search window over several parts of the input image, usually running over the whole image from the upper left to the lower right corner. To find objects on larger scales, the detector is enlarged by rescaling the features. This is effectively done by several look-ups in the integral image. In our approach, the search windows are only applied in the neighborhood of the region of interest detected by the attentional system.

5   Experiments and Results

We investigate the performance of the system on the example of finding chairs in an office environment. However, the future goal will be the construction of a flexible vision system that is able to search for and detect different object classes while the robot drives through its environment. If the robot moves, the time for the recognition is limited and it is not possible to search for many objects in a scene. A naive approach to restrict processing is to search the whole image for the first object class, then for the second class and so on. The problem is that if there is not enough time to check all object classes, some of the classes of the data base would never be checked. In our approach, we restrict processing to the salient regions in the image recognizing objects in order of their saliency.

To show the performance of the system, we claim three points: Firstly, the attention system detects regions of interest. Secondly, the classifier has good detection and false alarm rates on laser data. And finally, the combination of both systems yields a significant speed up and reliably detects objects at regions of interest. These three points will be investigated in the following.

Firstly, the performance of attention systems on camera data was evaluated by Parkhurst et al. [20] and Ouerhani et al. [21]. They demonstrate that attention systems based on the Koch-Ullman model [8] detect salient regions with a performance comparable to humans. We showed in [16] and [13] that attentional mechanisms work also reliably on laser data and that the two laser modes complement each other, enabling the consideration of more object qualities. Two examples of these results are depicted in Fig. 7.

Secondly, we tested the performance of the classifier. Its high performance for face detection was shown in [2], here we show the performance on laser data. The classifier was trained on laser images (300×300 pixels) of office chairs. We rendered 200 training images with chairs from 46 scans. Additionally, we provided 738 negative example images to the classifier from which a multiple of sub-images is created automatically.

The cascade in Fig. 6 presents the first three stages of the classifier for the object class "office chair" using depth values. One main feature is the horizontal bar in the first stage representing the seat of the chair. The detection starts with a classifier of size 20×40 pixels. To test the general performance of the classifier, the image is searched from top left to bottom right by applying the cascade. To detect objects at larger scales, the detector is rescaled. The classification is performed on a joint cascade of range and reflectance data. A logical "and" combines the results of both modes, yielding a reduction of false detections.

Fig. 7. Two examples of foci found by the attention system on laser data. Left: Camera image of the scene. Right: The corresponding scene in laser data with foci of attention. The traffic sign, the handicapped person sign and the person were focused (taken from our results in [13])

Fig. 8. The combination of both laser modes for classification reduces the amount of false detections. The false detection in the reflectance image (left) does not exist in the depth image (middle) and therefore it is eliminated in the combined result (right)

Fig. 8 shows an example of a recognized chair in a region found by the attention system. The false detection in the reflectance image (left) does not exist in the depth image (middle) and therefore is eliminated in the combined result (right). Table 1 summarizes the results of exhaustive classification, i.e., searching the whole image, with a test data set of 31 scans that are not used for learning (see also [22]). It shows that the number of false detections is reduced to zero by the combination of the modes while the detection rates change only slightly.

Finally, we show the results of the combination of attention and classification system and analyze the time performance. The coordinates of the focus serve as input for the classifier. Since a focus is not always at the center of an object but often at the borders, the classifier searches for objects in a specified region around the focus (here: radius 20 pixels). In this region, the classifier begins its search for objects with a 20×40 search window. To find chairs at larger scales, the detector is enlarged by rescaling the simple features.

Table 1. Detections and false detections of the classifier applied to 31 chair images

object class	no. of obj.	detections			false detections		
		refl. im.	depth im.	**comb.**	refl. im.	depth im.	**comb.**
chair	33	30	29	**29**	2	2	**0**

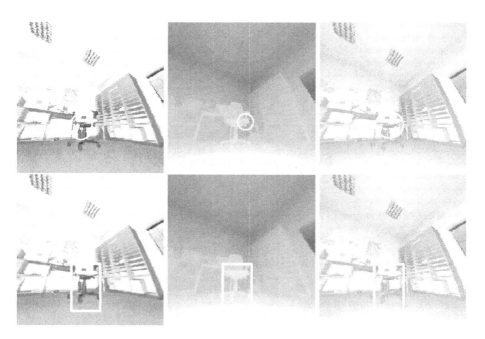

Fig. 9. A complete run of the recognition system. Top row: Most salient region in reflectance, depth, and combined image. The last one is transferred to the classifier. Bottom row: chair detected by the classifier in reflectance image, depth image and combined image

In all of our examples, the objects were detected if a focus of attention pointed to them and if the object was detected when searching the whole image. If no focus points to an object, this object is not detected. This is conform to our goal to detect only salient objects in the order of decreasing saliency.

Fig. 9 shows some images from a complete run of the recognition system. The top row depicts the most salient regions in the single reflectance (left) and depth image (middle) and the computed focus in the combination of both modes (right). The bottom row shows the rectangle that denotes a detected chair in reflectance and depth image and on the right the chair that is finally recognized by the joint cascade. Fig. 10 shows some more examples: the chairs are successfully detected even if the focus is at the object's border (left and middle) and if the object is partially occluded (right). However, severely occluded objects are not

Fig. 10. Three examples of chair detections. Top: Combined foci of attention. Bottom: corresponding detections by the classifier. Left: Chair is detected even if the focus is at its border. Middle: Detection of two chairs. Right: Chair is detected although it is presented sidewards and partially occluded

detected; the amount of occlusion still enabling detection has to be investigated further.

The classification needs 60 ms if a focus is provided as a starting point, compared to 200 ms for an uninformed search across the whole image (Pentium-IV-2400). So the focused classification needs only 30% of the time of the exhaustive one. The attention system requires 230 ms to compute a focus for both modes, i.e., for m object classes the exhaustive search needs $m * 200$ ms, the attentive search needs $230 + m * 60$ ms. Therefore, already for two different object classes in the data base, the return of investment is reached and the time saving increases proportionally with the number of objects.

6   Conclusions

In this paper, we have presented a new architecture for combining biologically motivated attentional mechanisms with a fast method for object classification. Input data are provided by a 3D laser scanner mounted on top of an autonomous robot. The scanner provides illumination-independent, bi-modal data that are transformed to depth and reflectance images. These serve as input to an attention system, directing the focus of attention sequentially to regions of potential interest. The foci determine starting regions for a cascade of classifiers which use

Haar-like features. By concentrating classification on salient regions, the classifier has to consider only a fraction of the search windows than in the case of an exhaustive search over the whole image. This speeds up the classification part significantly.

The architecture benefits from the fusion of the two laser modes resulting in more detected objects and a lower false classification rate. The range data enables the detection of objects with the same intensity like their background whereas the reflection data is able to detect flat objects. Moreover, misclassifications of shadows, mirroring objects and pictures of objects on the wall are avoided.

In future work, we will include top-down mechanisms in the attention model, enabling goal dependent search for objects. Furthermore, we plan to additionally integrate camera data into the system, allowing the simultaneous use of color, depth, and reflectance. The classifier will be trained for additional objects which compete for saliency. The overall goal will be a flexible vision system that recognizes salient objects first, guided by attentional mechanisms, and registers the recognized objects in semantic maps autonomously built by a mobile robot. The maps will serve as interface between robot and humans.

Acknowledgements. Special thanks to Joachim Hertzberg and Erich Rome for supporting our work.

References

1. Bennamoun, M., Mamic, G.: Object Recognition: Fundamentals and Case Studies. Springer (2002)
2. Viola, P., Jones, M.J.: Robust real-time face detection. International Journal of Computer Vision **57** (2004) 137–154
3. Treisman, A., Gelade, G.: A feature integration theory of attention. Cognitive Psychology **12** (1980) 97–136
4. Itti, L., Koch, C., Niebur, E.: A model of saliency-based visual attention for rapid scene analysis. IEEE Trans. on Pattern Analysis & Machine Intelligence **20** (1998) 1254–1259
5. Tsotsos, J.K., Culhane, S.M., Wai, W.Y.K., Lai, Y., Davis, N., Nuflo, F.: Modeling visual attention via selective tuning. AI **78** (1995) 507–545
6. Backer, G., Mertsching, B., Bollmann, M.: Data- and model-driven gaze control for an active-vision system. IEEE Trans. on Pattern Analysis & Machine Intelligence **23(12)** (2001) 1415–1429
7. Surmann, H., Lingemann, K., Nüchter, A., Hertzberg, J.: A 3d laser range finder for autonomous mobile robots. In: Proc. 32nd Intl. Symp. on Robotics (ISR 2001) (April 19–21, 2001, Seoul, South Korea). (2001) 153–158
8. Koch, C., Ullman, S.: Shifts in selective visual attention: towards the underlying neural circuitry. Human Neurobiology (1985) 219–227
9. Pessoa, L., Exel, S.: Attentional strategies for object recognition. In Mira, J., Sachez-Andres, J., eds.: Proc. of the IWANN, Alicante, Spain 1999. Volume 1606 of Lecture Notes in Computer Science., Springer (1999) 850–859
10. Miau, F., Papageorgiou, C., Itti, L.: Neuromorphic algorithms for computer vision and attention. In: Proc. SPIE 46 Annual International Symposium on Optical Science and Technology. Volume 4479. (2001) 12–23

11. Johnson, A., Hebert, M.: Using spin images for efficient object recognition in cluttered 3D scenes. IEEE Trans. on Pattern Analysis & Machine Intelligence **21** (1999) 433–449

12. Ruiz-Correa, S., Shapiro, L.G., Meila, M.: A New Paradigm for Recognizing 3-D Object Shapes from Range Data. In: Proc. Int'l Conf. on Computer Vision (ICCV '03), Nice, France (2003)

13. Frintrop, S., Rome, E., Nüchter, A., Surmann, H.: A bimodal laser-based attention system. (submitted)

14. Draper, B., Lionelle, A.: Evaluation of selective attention under similarity transforms. In: Proc. of the Int'l Workshop on Attention and Performance in Computer Vision (WAPCV'03), Graz, Austria (2003) 31–38

15. Lee, K., Buxton, H., Feng, J.: Selective attention for cue-guided search using a spiking neural network. In: Proc. of the Int'l Workshop on Attention and Performance in Computer Vision (WAPCV'03), Graz, Austria (2003) 55–62

16. Frintrop, S., Rome, E., Nüchter, A., Surmann, H.: An Attentive, Multi-modal Laser "Eye". In Crowley, J., Piater, J., Vincze, M., Paletta, L., eds.: Proc. of 3rd Int'l Conf. on Computer Vision Systems (ICVS 2003), Springer, Berlin, LNCS 2626 (2003) 202–211

17. Papageorgiou, C., Oren, M., Poggio, T.: A general framework for object detection. In: Proc. 6th Int'l Conf. on Computer Vision (ICCV '98), Bombay, India (1998) 555–562

18. Lienhart, R., Kuranov, A., Pisarevsky, V.: Empirical Analysis of Detection Cascades of Boosted Classifiers for Rapid Object Detection. In: Proc. 25th German Pattern Recognition Symposium (DAGM '03), Magdeburg, Germany (2003)

19. Freund, Y., Schapire, R.E.: Experiments with a new boosting algorithm. In: Machine Learning: Proc. 13th International Conference. (1996) 148–156

20. Parkhurst, D., Law, K., Niebur, E.: Modeling the role of salience in the allocation of overt visual attention. Vision Research **42** (2002) 107–123

21. Ouerhani, N., von Wartburg, R., Hügli, H., Müri, R.: Empirical validation of the saliency-based model of visual attention. Elec. Letters on Computer Vision and Image Analysis **3** (2004) 13–24

22. Nüchter, A., Surmann, H., Hertzberg, J.: Automatic Classification of Objects in 3D Laser Range Scans. In: Proc. 8th Conf. on Intelligent Autonomous Systems (IAS '04), Amsterdam, The Netherlands, IOS Press (2004) 963–970

A Visual Attention-Based Approach for Automatic Landmark Selection and Recognition

Nabil Ouerhani[1], Heinz Hügli[1], Gabriel Gruener[2], and Alain Codourey[2]

[1] Institute of Microtechnology, University of Neuchâtel,
Rue A.-L. Breguet 2, CH-2000 Neuchâtel, Switzerland
Nabil.Ouerhani@unine.ch
[2] Centre Suisse d'Electronique et Microtechnique, CSEM, Microrobotics Division,
Untere Gründlistrasse 1, CH-6055 Alpnach Dorf, Switzerland
Gabriel.Gruener@csem.ch

Abstract. Visual attention refers to the ability of a vision system to rapidly detect visually salient locations in a given scene. On the other hand, the selection of robust visual landmarks of an environment represents a cornerstone of reliable vision-based robot navigation systems. Indeed, can salient scene locations provided by visual attention be useful for robot navigation? This work investigates the potential and effectiveness of the visual attention mechanism to provide pre-attentive scene information to a robot navigation system. The basic idea is to detect and track the salient locations, or spots of attention by building trajectories that memorize the spatial and temporal evolution of these spots. Then, a persistency test, which is based on the examination of the lengths of built trajectories, allows the selection of good environment landmarks. The selected landmarks can be used for feature-based localization and mapping systems which helps mobile robot to accomplish navigation tasks.

1 Introduction

Visual attention is the natural ability of the human visual system to quickly select within a given scene specific parts deemed important or salient by the observer. In computer vision, a similar visual attention mechanism designates the first low-level processing step that allows to quickly selecting in a scene the points of interest to be analyzed more specifically and in-depth in a second processing step.

The computational modeling of visual attention has been a key issue in artificial vision during the last two decades [1, 2, 3]. First reported in 1985 [4], the saliency-based model of visual attention is largely accepted today [5] and gave rise to numerous soft and hardware implementations [5, 6]. In addition, this model has been used in several computer vision applications including image compression [7] and color image segmentation [8].

In visual robot navigation, the detection, tracking, and selection of robust visual landmarks represent the most challenging issues in building reliable

L. Paletta et al. (Eds.): WAPCV 2004, LNCS 3368, pp. 183–195, 2005.

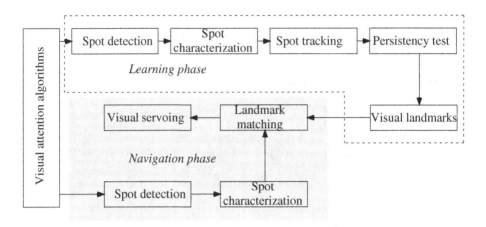

Fig. 1. Overview of the attention-based landmark selection approach

navigation systems [9, 10]. Numerous previous works have pointed to the visual
attention paradigm in solving various issues in active vision in general [11, 12]
and visual robot navigation in particular [13].

This work proposes a visual attention-based approach for visual landmark
selection. The proposed approach relies on an extended version of Itti's *et al.*
model of visual attention [5] in order to detect the most visually salient scene
locations; the spots of attention. More specifically, these spots of attention are
deduced from a saliency map computed from multiple visual cues including corner features. Then, the spots of attention are characterized using a feature vector
that represents the contribution of each considered feature to the final saliency of
the spot. Once characterized, the spots of attention are easily tracked over time
using a simple tracking method that is based on feature matching. The tracking
results reveal the persistency and thus the robustness of the spots, leading to a
reliable criterium for the selection of the landmarks.

The navigation phase, which has not been fully tested yet consists in using the
selected environment landmarks for feature-based Simultaneous Localization and
Mapping (SLAM) on a mobile robot [14]. A schematic overview of the landmark
selection approach as well as its integration into a general visual robot navigation
system are given in Figure 1.

The remainder of this paper is organized as follows. Section 2 describes the
saliency-based model of visual attention. Section 3 presents the characterization
and tracking of spots of attention. The persistency test procedure that allows
the selection and the representation of the landmarks is exposed in Section 4. In
Section 5, a landmark recognition method is described. Section 6 reports some
experiments carried out on real robot navigation image sequences in order to
assess the approach proposed in this paper. Finally, the conclusions and some
perspectives are stated in Section 7.

2   Attention-Based Landmark Detection

2.1   Saliency-Based Model of Visual Attention

The saliency-based model of visual attention, which selects the most salient parts of a scene, is composed of four main steps [4, 5].

1) First, a number of features are extracted from the scene by computing the so called feature maps F_j. The features most used in previous works are intensity, color, and orientation. The use of these features is motivated by psychophysical studies on primate visual systems. In particular, the authors of the model used two chromatic features that are inspired from human vision, namely the two opponent colors red/green (RG) and blue/yellow (BY).

2) In a second step, each feature map F_j is transformed in its conspicuity map C_j. Each conspicuity map highlights the parts of the scene that strongly differ, according to a specific feature, from its surrounding. This is usually achieved by using a *center-surround*-mechanism which can be implemented with multiscale *difference-of-Gaussian*-filters. It is noteworthy that this kind of filters have been used by D. Lowe for extracting robust and scale-invariant features (SIFT) from grey-scale images for object recognition, stereo matching but also for robot navigation [10, 15].

3) In the third stage of the attention model, the conspicuity maps are integrated together, in a competitive way, into a *saliency map* $\mathcal{S}$ in accordance with equation 1.

$$\mathcal{S} = \sum_{j=1}^{J} \mathcal{N}(\mathcal{C}_j) \tag{1}$$

where $\mathcal{N}()$ is a normalization operator that promotes conspicuity maps in which a small number of strong peaks of activity are present and demotes maps that contain numerous comparable peak responses [5].

4) Finally the most salient parts of the scene are derived from the saliency map by selecting the most active locations of that map. A Winner-Take-All network (WTA) is often used to implement this step [4].

2.2   Extension of the Model to Corner Features

In the context of vision-based robot navigation, corner features are considered as highly significant landmark candidates in the navigation environment [9, 16]. This section aims at extending the basic model of visual attention to consider also corner features. To do so, a corner map C_c which highlights the corner points in the scene, is first computed. Then, this corner map is combined together with the color and intensity-based conspicuity maps into the final saliency map.

Multi-Scale Harris Corner Detector [17, 18]. Practically, the proposed multiscale method computes a corner pyramid $\mathcal{P}_c$. Each level of the corner pyramid detects corner points at a different scale. Formally, $\mathcal{P}_c$ is defined according to Equation 2.

$$\mathcal{P}_c(i) = Harris(\mathcal{P}_g(i)) \tag{2}$$

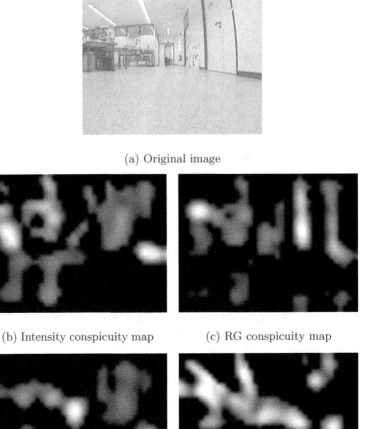

(a) Original image

(b) Intensity conspicuity map   (c) RG conspicuity map

(d) BY conspicuity map   (e) Corner conspicuity map

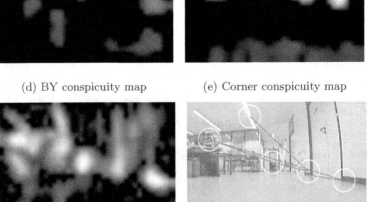

(f) Saliency map   (g) Spots of attention

Fig. 2. Example of the Conspicuity maps, the saliency map and the corresponding spots of attention computed with the corner-extended model of visual attention

where $Harris(.)$ is the Harris corner detector as defined in [17] and $\mathcal{P}_g$ is a gaussian pyramid defined as follows [19]:

$$\mathcal{P}_g(0) = I$$
$$\mathcal{P}_g(i) = \left(\downarrow 2 \right) (\mathcal{P}_g(i-1) * G) \tag{3}$$

where I is a grey-scale version of the input image, G is a gaussian filter and $\left(\downarrow 2 \right)$ refers to the down-sampling (by 2) operator.

Corner Conspicuity Map C_c. Given the corner pyramid $\mathcal{P}_c$, C_c is computed in accordance with Equation 4.

$$C_c = \sum_{s=1}^{s_{max}} \mathcal{P}_c(s) \tag{4}$$

Note that the summation of the multiscale corner maps $\mathcal{P}_c(s)$ is achieved at the coarsest resolution. Maps of finer resolutions are lowpass filtered and downsampled to the required resolution. In our implementation s_{max} is set to 4, in order to get a corner conspicuity map C_c that has the same resolution as the color- and intensity-related conspicuity maps.

Integration of Corner Feature into the Model. The final saliency map $\mathcal{S}$ of the extended model is computed in accordance with Equation 5.

$$\mathcal{S} = \sum_{j=1}^{J+1} \mathcal{N}(C_j) \tag{5}$$

where

$$C_{J+1} = C_c \tag{6}$$

Selection of the Spots of Attention. The maxima of the saliency map represent the most salient spots of attention. Once a spot is selected, a region around its location is inhibited in order to allow the next most salient spot to be selected. The total number of spots of attention can be either set interactively or automatically determined by the activity of the saliency map. For simplicity, the number of spots is set to five in our implementation.

Figure 2 shows an example of the four conspicuity maps, saliency map and the spots of attention computed by the corner-extended model of visual attention.

3   Spot Characterization and Tracking

3.1   Spot Characterization

The spots of attention computed by means of the extended model of visual attention locate the scene features to be tracked. In addition to location, each spot $\mathbf{x}$ is also characterized by a feature vector $\mathbf{f}$:

$$\mathbf{f} = \begin{pmatrix} f_1 \\ .. \\ f_J \end{pmatrix} \tag{7}$$

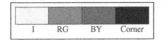

(a) Feature represen-
tation

 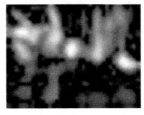 

(b) Original image   (c) Saliency map   (d)   Characterized
                                                      spots of attention

Fig. 3. Characterization of spots of attention. The five most salient spots of attention are detected and characterized using four visual features, namely intensity (I), red-green (RG) and blue-yellow (BY) color components, and corners

where J is the number of the considered features in the attention model and f_j refers to the contribution of the feature j to the detection of the spot $\mathbf{x}$. Formally, f_j is computed as follows:

$$f_j = \frac{\mathcal{N}(C_j(\mathbf{x}))}{\mathcal{S}(\mathbf{x})} \qquad (8)$$

Note that $\sum_{j=1}^{J}(f_j) = 1$.

Let N be the number of frames of a sequence and M the number of spots detected per frame, the spots of attention can be formally described as $P_{m,n} = (\mathbf{x}_{m,n}, \mathbf{f}_{m,n})$, where $m \in [1..M]$, $n \in [1..N]$, $\mathbf{x}_{m,n}$ is the spatial location of the spot, and $\mathbf{f}_{m,n}$ its characteristic feature vector. Figure 3 illustrates an example of the characterization of spots of attention.

3.2   Spot Tracking

The basic idea behind the proposed algorithm is to build a trajectory for each tracked spot of attention. Each point of the trajectory memorizes the spatial and the feature-based information of the tracked spot at a given time.

Specifically, given the M spots of attention computed from the first frame, the tracking algorithm starts with creating M initial trajectories, each of which contains one of the M initial spots. The initial spots represent also the head elements of the initial trajectories. A new detected spot $P_{m,n}$ is either appended to an existing trajectory (and becomes the head of that trajectory) or gives rise to a new trajectory, depending on its similarity with the head elements P^h of already existing trajectories as described in Algorithm 1. Note that a spot of attention is assigned to exactly one trajectory (see the parameter *marked[]* in Algorithm 1) and a trajectory can contain at most one spot from the same frame. In a simple implementation, the condition that a spot $P_{m,n}$ must fulfil in order to be appended to a trajectory T with a head element $P^h = (\mathbf{x}_h, \mathbf{f}_h)$ is given by:

$$P_{m,n} \in T \; \; if \; \; \|\mathbf{x}_{m,n} - \mathbf{x}_h\| < \epsilon_{\mathbf{x}} \; \& \; \|\mathbf{f}_{m,n} - \mathbf{f}_h\| < \epsilon_{\mathbf{f}} \tag{9}$$

where $\epsilon_{\mathbf{x}}$ and $\epsilon_{\mathbf{f}}$ can be either determined empirically or learned from a set of image sequences.

In the absence of ground-truth data, the evaluation of the tracking algorithm can be achieved interactively. Indeed, a human observer can visually judge the correctness of the trajectories, i.e. if they track the same physical scene constituents. Figure 4 gives some examples of trajectories built from a set of spots of attention using the tracking algorithm described above.

In a more advanced version of the tracking algorithm, Kalman filter is expected to enhance the tracking performance. Indeed, in the presence of different sources of information such as images and odometry, Kalman filter becomes an intuitive framework for tracking. In addition, the predictive nature of the filter decreases the probability of feature loss during tracking.

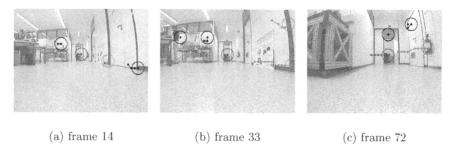

(a) frame 14   (b) frame 33   (c) frame 72

Fig. 4. Examples of trajectories built from a set of spots of attention

4   Landmark Selection and Representation

This step of the approach is part of the learning phase and aims at selecting, among all detected spots of attention, the most robust as visual landmarks of the environment. The basic idea is to examine the trajectories built while tracking spots of attention. Specifically, the length of the trajectories reveals the robustness of the detected spots of attention. Thus, during the learning phase the cardinality $(Card(T))$ of a trajectory directly determines whether the corresponding spots of attention are good landmarks. Thus, a landmark L is created for each trajectory T_L that satisfies the described robustness criterium.

In addition, the cardinality of the trajectories can be used as measure to compare the performance of different interest points detectors, as stated in Section 6, but also of different tracking approaches.

Algorithm 1 Attention-based object tracking

Image sequence $I(n)$ $(1..n..N)$
Number of detected spots of attention per frame: M
Boolean *appended*
Boolean *marked*[ ]
Trajectory set $\{T\} = \emptyset$

for $n = 1..N$ **do**
    Detect & characterize the M spots of attention $P_{m,n} = (\mathbf{x}_{m,n}, \mathbf{f}_{m,n})$
    for $k = 1..card(\{T\})$ **do**
      $marked[k] = 0$
    end for
    for $m = 1..M$ **do**
      $appended = 0$
      for $k = 1..card(\{T\})$ **do**
        if $(marked[k] == 0)$ **then**
          if $d(P_{m,n}, P_k^h) < \varepsilon$ * **then**
            $append(P_{m,n}, T_k)$
            $appended = 1$
            $marked[k] = 1$
            break
          end if
        end if
      end for
      if $(appended == 0)$ **then**
        $newTraject(T_{card(\{T\})+1})$
        $append(P_{m,n}, T_{card(\{T\})+1})$
        $\{T\} = \{T\} \cup \{T_{card(\{T\})+1}\}$
      end if
    end for
end for
 * $d()$ is given by Equation 9

Once selected, the landmarks should be then represented in an appropriate manner in order to best describe the navigation environment. In this work, two attributes are assigned to each landmark L: spatial attribute and feature-based attribute. Regarding the spatial attribute, the height of the scene constituents is constant since the robot is navigating on flat ground and the camera is fixed on the robot. Thus, the y-coordinate of the selected landmarks is independent of the robot orientation and varies only slightly. Therefore, the y-coordinates y_L as well as its maximum variation Δy_L are considered as landmark attributes. For the feature-based attributes, the mean feature vector μ_L of all spots belonging to the landmark-related trajectory as well as its standard deviation σ_L are the two attributes assigned to a landmark L. μ_L and σ_L are defined in accordance with Equation 10.

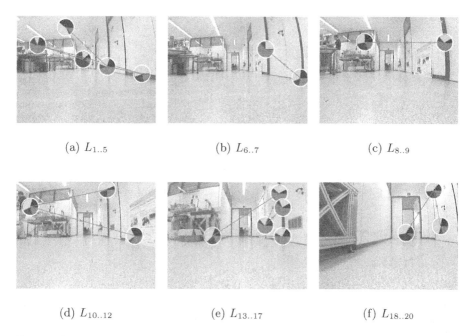

(a) $L_{1..5}$ (b) $L_{6..7}$ (c) $L_{8..9}$

(d) $L_{10..12}$ (e) $L_{13..17}$ (f) $L_{18..20}$

Fig. 5. The selected landmarks and their attributes computed from sequence 1. The red arrow indicates the increasing index of the landmarks L_i

$$\mu_L = \frac{1}{Card(T_L)} \sum_{m,n|P_{m,n} \in T_L} (\mathbf{f}_{m,n})$$

$$\sigma_L = \sqrt{\frac{1}{Card(T_L)} \sum_{m,n|P_{m,n} \in T_L} (\mathbf{f}_{m,n} - \mu_L)^2} \qquad (10)$$

To summarize, a landmark L is described by a four component vector $(y_L, \Delta y_L, \mu_L, \sigma_L)^T$. Figure 6 shows the landmarks that have been automatically selected and represented from a lab navigation environment.

5   Landmark Recognition

During navigation, a robot has to detect and identify previously learned landmarks in order to localize itself in the environment. In this work, we propose a landmark recognition method that relies on characterized spot matching similar to the one described in Section 3. Specifically, given a set of landmarks $L_i(y_{L_i}, \Delta y_L, \mu_{L_i}, \sigma_{L_i}))$ learned during the exploration phase and a detected spot of attention $P_{m,n}(\mathbf{x}_{m,n}, \mathbf{f}_{m,n})$ (with $\mathbf{x}_{m,n} = (x_{m,n}, y_{m,n})$), then the landmarks that correspond to this spot are those L_i that satisfy the following criteria:

$$|y_{L_i} - y_{m,n}| \leq k \cdot \Delta y_L \quad \&\& \qquad (11)$$
$$\|\mu_{L_i} - \mathbf{f}_{m,n}\| \leq k \cdot \sigma_{L_i}$$

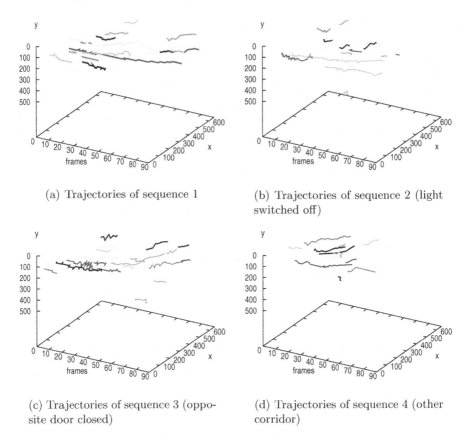

(a) Trajectories of sequence 1

(b) Trajectories of sequence 2 (light switched off)

(c) Trajectories of sequence 3 (opposite door closed)

(d) Trajectories of sequence 4 (other corridor)

Fig. 6. Trajectories of the tracked spots of attention from four different sequences ((a)..(d)). Note that only trajectories with $Card(T) > 3$ are represented here

where k is a control parameter for the tolerance/strictness of the matching. Setting $k \in [1.5 .. 2]$ leads to satisfying results. Note that our recognition method uses a soft matching scheme, e.i. a spot of attention can be matched to more than one landmark.

In future work, we are intending to exploit the spatial relationships between single spots but also between landmarks, in order to remove false matchings. For instance, the spatial order constraints of landmarks presented in [20] is a possible solution for this problem.

6   Results

This section presents some experiments that aim at assessing the presented landmark selection approach. The tests have been carried out with four sequences acquired by a camera mounted on a robot that navigates in an indoor environment over a distance of about 10 meters (see Figure 3). The length of the sequences varies between 60 and 83 frames. Two groups of results are presented here. Qualitative results regarding the robustness of the detection and tracking algorithms and quantitative results that point to the superiority of the corner-extended model of attention over the classic one.

Regarding the first group of results, Figure 6 illustrates the trajectories built from each sequence. The trajectories are plotted in 3D (x, y, t) in order to better visualize their temporal extent.

In the first sequence (Figure 6(a)), the most robustly detected and tracked landmark is the entrance of the differently illuminated room toward which the robot is moving. The trajectory built around this landmark has a length of 83, which means that the spot has been detected in each frame of the sequence. In addition, the red-colored door frames (especially their corners) and a fire extinguisher have been tracked over a large number of frames. Ceiling lights figure also between the detected and tracked features.

Like the first example, the three others ((b) light switched off, (c) front door closed, and (d) other corridor) tend to show, qualitatively, the ability of the proposed approach to robustly detect and track certain visual features of the navigation environment over a large period of time and under different conditions. For instance, the door frames and the fire extinguisher figure among those features that can be considered as environment landmarks. A more quantitative and in-depth evaluation of the robustness of the proposed approach towards view angle changes and changing in lighting conditions is required, in order to definitely validate our landmark selection method.

Table 1. Impact of the integration of Harris corner features on the tracking algorithm. The total number of trajectories, the minimum, maximum, and mean cardinality of trajectories are computed for the classical (without Harris) and the corner-extended (with Harris) models

	number of spots	number of trajectories		Min / Max($Card(T)$)		Mean($Card(T)$)	
		no Harris	Harris	no Harris	Harris	no Harris	Harris
Seq1	415	88	58	1 / 68	1 / 83	4.7	7.1
Seq2	320	136	80	1 / 19	1 / 56	2.3	4.0
Seq3	385	130	61	1 / 22	1 / 48	2.9	6.5
Seq4	280	123	56	1 / 22	1 / 31	2.2	5.0

Table 1, which resumes the second group of results, shows the advantage of the corner-extended model over the basic model regarding the stability of the detected spots of attention over time. For each of the four image sequences the total number of trajectories, their minimum, maximum, and mean cardinality (length) are represented. It can be seen that the integration of the corner features has leaded to more consistent trajectories.

7   Conclusions and Future Work

This work presents an attention-based approach for selecting visual landmarks in a robot navigation environment. An extended version of the saliency-based model of visual attention that considers also corners has been used to extract spatial and feature-based information about the most visually salient locations of a scene. These locations are then tracked over time. Finally, the most robustly tracked locations are selected as environment landmarks. One of the advantages of this approach is the use of a multi-featured visual input, which allows to cope with navigation environments of different natures, while preserving, thanks to the feature competition, a discriminative characterization of the potential landmarks. Qualitative results show the ability of the method to select good environment landmarks, whereas the quantitative results confirm the superiority of the corner-extended model of attention over the classic one, regarding the consistency of the detected spots of attention over time.

In future work, the rather simple tracking algorithm will be improved, essentially by introducing predictive filters such as Kalman and particle filters [21]. A quantitative evaluation of the landmark recognition method is one of the next steps to be done. In addition, we are planning to apply the proposed approach to solve some problems related to Simultaneous Localization and Map building (SLAM) in real robot navigation tasks.

References

1. B. Julesz and J. Bergen. Textons, the fundamental elements in preattentive vision and perception of textures. *Bell System Technical Journal, Vol. 62, No. 6, pp. 1619-1645*, 1983.
2. J.M. Wolfe. Guided search 2.0: A revised model of visual search. *Psychonomic Bulletin & Review, Vol. 1, pp. 202-238*, 1994.
3. J.K. Tsotsos, S.M. Culhane, W.Y.K. Wai, Y.H. Lai, N. Davis, and F. Nuflo. Modeling visual attention via selective tuning. *Artificial Intelligence, Vol. 78, pp. 507-545*, 1995.
4. Ch. Koch and S. Ullman. Shifts in selective visual attention: Towards the underlying neural circuitry. *Human Neurobiology, Vol. 4, pp. 219-227*, 1985.
5. L. Itti, Ch. Koch, and E. Niebur. A model of saliency-based visual attention for rapid scene analysis. *IEEE Transactions on Pattern Analysis and Machine Intelligence (PAMI), Vol. 20, No. 11, pp. 1254-1259*, 1998.
6. N. Ouerhani and H. Hugli. Real-time visual attention on a massively parallel SIMD architecture. *International Journal of Real Time Imaging, Vol. 9, No. 3, pp. 189-196*, 2003.
7. N. Ouerhani, J. Bracamonte, H. Hugli, M. Ansorge, and F. Pellandini. Adaptive color image compression based on visual attention. *International Conference on Image Analysis and Processing (ICIAP'01), IEEE Computer Society Press, pp. 416-421*, 2001.
8. N. Ouerhani and H. Hugli. Maps: Multiscale attention-based presegmentation of color images. *4th International Conference on Scale-Space theories in Computer Vision, Springer Verlag, Lecture Notes in Computer Science (LNCS), Vol. 2695, pp. 537-549*, 2003.

9. A.J. Davison. *Mobile Robot Navigation Using Active Vision.* PhD thesis, University of Oxford, UK, 1999.
10. S. Se, D. Lowe, and J. Little. Global localization using distinctive visual features. *International Conference on Intelligent Robots and Systems, IROS, pp. 226-231,* 2002.
11. J.J. Clark and N.J. Ferrier. Control of visual attention in mobile robots. *IEEE Conference on Robotics and Automation, pp. 826-831,* 1989.
12. K. Brunnstrom, J.O. Eklundh, and T. Uhlin. Active fixation for scene exploration. *International Journal of Computer Vision, Vol. 17, pp. 137-162,* 1994.
13. L. Itti. Toward highly capable neuromorphic autonomous robots: beobots. *SPIE 47 Annual International Symposium on Optical Science and Technology, Vol. 4787, pp. 37-45,* 2002.
14. M.W.M. Dissanayakeand Gamini, P. Newman, S. Clark, H.F. Durrant-Whyte, and M. Csorba. A solution to the simultaneous localization and map building (slam) problem. *IEEE Transactions on Robotics and Automation, Vol. 17, pp. 229-241,* 2001.
15. D. Lowe. Distinctive image features from scale-invariant keypoints. *International Journal of Computer Vision, Vol. 60 (2), pp. 91-110,* 2004.
16. A.A. Argyros, C. Bekris, and S. Orphanoudakis. Robot homing based on corner tracking in a sequence of panoramic images. *Computer Vision and Pattern Recognition Conference (CVPR), pp. 11-13,* 2001.
17. C.G. Harris and M. Stephens. A combined corner and edge detector. *Fourth Alvey Vision Conference, pp. 147-151,* 1988.
18. K. Mikolajczyk and C. Schmid. An affine invariant interest point detector. *European Conference on Computer Vision (ECCV), Vol.1, pp. 128-142,* 2002.
19. P.J. Burt and E.H. Adelson. The laplacian pyramid as a compact image code. *IEEE Transactions on Communication Vol. 31 (4) pp. 532-540,* 1983.
20. A. Tapus, N. Tomatis, and R. Siegwart. Topological global localization and mapping with fingerprint and uncertainty. *International Symposium on Experimental Robotics,* 2004.
21. D. Fox, S. Thrun, F. Dellaert, and W. Burgard. Particle filters for mobile robot localization. *In A. Doucet, N. de Freitas, and N. Gordon, editors, Sequential Monte-Carlo Methods in Practice. Springer-Verlag, New York,* 2000.

Biologically Motivated Visual Selective Attention for Face Localization

Sang-Woo Ban and Minho Lee

School of Electronic and Electrical Engineering,
Kyungpook National University
1370 Sankyuk-Dong, Puk-Gu, Taegu 702-701, Korea
swban@palgong.knu.ac.kr, mholee@knu.ac.kr
http://abr.knu.ac.kr

Abstract. We propose a new biologically motivated model to localize or detect faces in natural color input scene. The proposed model integrates a bottom-up selective attention model and a top-down perception model. The bottom-up selective attention model using low level features sequentially selects a candidate area which is preferentially searched for face detection. The top-down perception model consists of a face spatial invariant feature detection model using ratio template matching method with training mechanism and a face color perception model, which is to model the roles of the inferior temporal areas and the V4 area, respectively. Finally, we construct a new face detection model by integration of the bottom-up saliency map model, the face color perception model and the face spatial invariant feature detection model. Computer experimental results show that the proposed model successfully indicates faces in natural scenes.

1 Introduction

Considering the human-like selective attention function, top-down or task dependent processing can affect how to determine the saliency map as well as bottom-up or task independent processing [1]. In the top-down manner, the human visual system determines salient locations through perceptive processing such as understanding and recognition. It is well known that the perception mechanism is one of the most complex activities in our brain. Moreover, top-down processing is so subjective that it is very difficult to model the processing mechanism in detail. On the other hand, with bottom-up processing, the human visual system determines salient locations obtained from features that are based on the basic information of an input image such as intensity, color, and orientation, etc. [1]. Bottom-up processing can be considered as a function of primitive selective attention in human vision system since humans selectively attend to such a salient area according to various stimuli in input scene. In the course of detecting a face, both the bottom-up and the top-down processing work together for selective attention of a face region.

In the last five years, face and facial expression recognition have attracted much attention though they have been studied for more than 20 years by psychophysicists, neuroscientists, and engineers [2]. Numerous methods have been developed to localize or detect faces in a visual scene. Yang et al, [2] have reviewed and classified those

L. Paletta et al. (Eds.): WAPCV 2004, LNCS 3368, pp. 196–205, 2005.
© Springer-Verlag Berlin Heidelberg 2005

face detection methods into four major categories such as the knowledge-based methods, the feature invariant approaches, the template matching methods, and the appearance-based methods. According to the survey, no specific method has yet shown comparable performance with a human being. However, we can easily indicate that most of the proposed methods do not fully consider the biological mechanism of a human visual system which can effortlessly detect a face in natural or cluttered scenes. In order to develop more efficient face detection system, we should consider a computational model to gain insight into biological brain function. Recently, biologically motivated approaches have been developed by Itti, Poggio and Koch [3, 4]. However, they have not shown plausible performance for face detection until now.

In this paper, we propose a new computational model to detect a face based on understating the biological visual attention mechanism. The proposed model uses both the bottom-up saliency map model and the top-down visual perception model for face detection. When we construct the bottom-up saliency map model, we consider the roles of the cells in retina, lateral geniculate nucleus (LGN) and visual cortex. As the primitive features of an input scene, intensity, edge, color opponent and symmetry information are used as inputs of the bottom-up saliency map model. The independent component analysis (ICA) is used for integration of the feature maps, which simple reflects the redundancy reduction function of the primary visual cortex [5, 6, 7]. For the top-down visual perception model for face detection, we consider the roles of face-form selective cells in the inferior temporal (IT) areas and those of the neurons with face-color ones in the V4 area that locate in secondary visual areas. We propose a face color perception model and a spatial invariant feature detection model using a ratio template matching method which is based on the constant property of relative intensity between sub-regions of face. In our model, the ratio template matching model is incorporated into a training mechanism implemented by auto associative multi-layer perceptron (AAMLP). Finally, we construct a face conspicuity map by a binding process that integrates the bottom-up saliency map model, the top-down ratio-template matching model and face-color perception model. The proposed model can tremendously reduce the face detecting time by considering not all the area of an input image but only the attention area decided by the bottom-up saliency map model.

Section 2 describes the bottom-up saliency map model. Section 3 deals with the top-down processing for face detection. Experimental results and discuss are shown in section 4. Conclusion and future works will be made in the end.

2   Bottom-Up Saliency Map Model

In the vertebrate retina, three types of cells are important processing elements for performing edge extraction [8, 9, 10]. On the other hand, a neural circuit in the retina creates color opponent coding such as R+G-, B+Y- , and B-W+ [11]. Those preprocessed signal transmitted to the LGN through the ganglion cell, and the on-set and off-surround mechanism of the LGN and the visual cortex intensifies the phenomena of opponency [11]. Moreover, the LGN and the primary visual cortex play a role of detecting a primitive shape or pattern of an object [11]. Even though the role of the visual cortex for finding a salient region is important, it is very difficult to model the detail function of the visual cortex. Fig. 1 (a) shows the biological visual pathway for face detection.

Fig. 1 (b) shows our bottom-up saliency map model. In order to model the human-like bottom-up visual attention mechanism, we used 4 bases of edge (E), intensity (I), color (RG and BY) and symmetry information (Sym) as shown in Fig. 1 (b), for which the roles of retina cells and lateral geniculate nucleus (LGN) are reflected in the previously proposed attention model [7]. The feature maps ($\overline{I}$, $\overline{E}$, $\overline{Sym}$, and $\overline{C}$) are constructed by center surround difference and normalization (CSD & N) of 4 bases, which mimics the on-center and off-surround mechanism in our brain, and then are integrated by an ICA algorithm [7]. The ICA can be used for modeling the roles of the primary visual cortex for redundancy reduction according to Barlow's hypothesis and Sejnowski's result [5, 6, 12, 13]. Barlow's hypothesis is that human visual cortical feature detectors might be the end result of a redundancy reduction process [5], and Sejnowski's result shows that the ICA is the best way to reduce redundancy [6, 12].

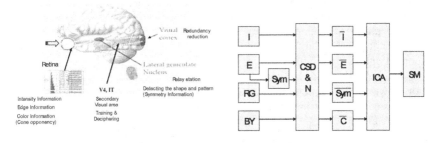

(a) Biological visual pathway for   (b) Arhitecture of the previsouly
        face detection   proposed model

Fig. 1. Biological visual pathway for face detection and the architecture of the previously proposed bottom-up saliency map model. (I : intensity feature, E : edge feature, Sym : symmetry feature, RG : red-green opponent coding feature, BY : blue-yellow opponent coding feature, CSD & N : center-surround difference and normalization, $\overline{I}$: intensity feature map, $\overline{E}$: edge feature map, $\overline{Sym}$: symmetry feature map, $\overline{C}$: color feature map, ICA : independent component analysis, SM : saliency map)

In our simulation, to obtain ICA filters, we derived the four characteristic maps ($\overline{I}$, $\overline{E}$, $\overline{Sym}$, and $\overline{C}$) that are used for input patches of the ICA. The basis functions are determined using the extended infomax algorithm [12]. Each basis function represents an independent filter, which is ordered according to the length of the filter vector. We apply the obtained ICA filters ICs_{ri} to the four characteristic maps FM_r ($\overline{I}$, $\overline{E}$, $\overline{Sym}$, and $\overline{C}$) as shown in Eq. (1), and obtain saliency map according to Eq. (2).

$$E_{ri} = FM_r * ICs_{ri} \qquad for \ i=1,...,N, \quad r=1,..,4 \tag{1}$$

$$S(x, y) = \sum E_{ri}(x, y) \quad for \ all \ i \tag{2}$$

where N denotes the number of filters. The convolution result, E_{ri} represents the influences of the four characteristic maps on each independent component. Then, the most salient point is selected by finding the location (x, y) with maximum summation of the values of all the pixels in the local window centered at the *(x,y)* in the saliency map. After we find the most salient point, we need to mask the previous salient point to find the next salient location, which models an inhibition of return (IOR) mechanism of the human vision system [7]. Fig. 2 shows the experimental result of the bottom-up saliency map model of a color image with faces.

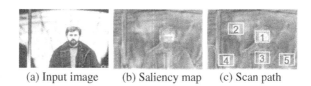

(a) Input image   (b) Saliency map   (c) Scan path

Fig. 2. Selective attention results of natural scenes. (The blocks in (c) represent selective attention areas, and each number in the blocks indicates the scan path order)

3 Top-Down Face Perception Model

It is well known that the cells in the IT area play an important role for detection and recognition of human faces using the processed visual information by the visual cortex. The face-selective cells in the IT area located in the secondary visual area contain complex shape coding information, and they generate a corresponding activity according to face shape information. Also, the neurons in the V4 area respond well to specific colors of objects, and the cell activities are irrespective of lighting conditions. The bottom-up processing conducts low-level feature extraction and the top down processing is related with very complex roles of our brain such as perception, emotion and memory, etc. It is not well known yet how the bottom-up visual processing mechanism integrates with the top-down visual perception mechanism for face detection. In our model, we modeled the binding process by simple summation operator. The proposed face detection model considered the bottom-up saliency map model as well as the face ratio template matching model and the face color perception model as shown in Fig. 3.

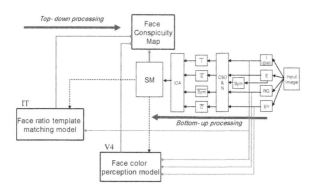

Fig. 3. The proposed face detection model

3.1 Face Detection Using Ratio Template Method

Human being has a perceptual constancy mechanism that is important not only for understanding color perception, but for understanding other types of perception as well. Perception constancy is that properties of objects remain constant in our perception process even when the conditions of stimulation are changed. Human vision perception has a lightness constancy that is our perception of an object's achromatic colors as remaining constant even when it is illuminated by lights with different intensities [11]. Human occasionally perceives two physically different areas as the same or almost same ones because they are on different backgrounds. We called the principle behind this effect as the ratio principle [11]. According to this principle, two areas that reflect different amounts of light will look the same if the ratios of their intensities to the intensities of the surrounds are kept constant. From this ratio principle, we can get the insights to detect a human face. Owing to the morphology of a human face, the relative degree of the brightness between sub-regions of a human face is perceived constantly while variations in illumination change the individual brightness of different parts of faces such as eyes, cheeks, mouse, chin, and forehead. Based on this mechanism, the ratio template for a face with a few appropriately chosen sub-regions is obtained from the face images. We considered the ratio template matching model as a part of the face form detection model because the ratio template for the face can be generated owing to the characteristic morphology of the face. Fig. 4 shows the ratio templates which are composed of 6 characteristic regions such as eyes, cheeks, mouse, chin, and forehead shown by small boxes and 6 relations shown by arrows. Each arrow in Fig. 4 indicates a contrast polarity relation between two sub-regions. In Fig. 4, the number in each region represents the average of intensity values of the each region.

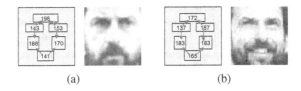

(a)   (b)

Fig. 4. Examples of ratio templates of face regions obtained from the intensity information of natural scenes. (Left column: ratio template, Right column: intensity image of face region)

Moreover, infant face perception shows rapid learning in the first few hours and days after birth through more experience and training [14]. In this paper, we used a training model to mimic such an infant face perception mechanism.

Although faces have tremendous within-class variability, face detection problem can be considered as a 2-class recognition problem, one of which is a face class and the other is a non-face class. Such a face detection problem might be one of the partially-exposed environments problems which are well known problems where training data from on-class is very little or non-existent. An Auto-associative neural network has been used successfully in many such partially-exposed environments [15]. In the proposed model, we use an AAMLP with 4-layers which are mapping layer, bottleneck layer, de-mapping layer and output layer. An auto-associative neural network is basically a neural network whose input and target vectors are the same

[15]. The training process of an AAMLP in the proposed model mimics the training mechanism of face-selective cells in the secondary visual area. The infants can not recognize well an unfamiliar faces at first, but after seeing and training many enough the infants can recognize their parents. In a course of test phase, the correlation of inputs and outputs of the AAMLP can imitate the activities of the face-selective cells in the inferotemporal area. Moreover, we don't need to have a non-face data set if we use the AAMLP for face detection.

Fig. 5 shows the proposed face detection model constructed by a 4-layer AAMLP. In general, we should consider the various conditions such as light luminescence, background image complexity and so on, in constructing the face detection system. In the proposed model, we use a ratio template represented by a 6 dimension ratio vector getting from a patch of face region in intensity feature image as the input data for the AAMLP. Let F denote an auto-associative mapping function, and x_i and y_i indicate an input and an output vector, respectively. Then the function F is usually trained to minimize the following mean square error given by Eq. (3).

$$E = \sum_{i=1}^{n} \| x_i - y_i \|^2 = \sum_{i=1}^{n} \| x_i - F(x_i) \|^2 \tag{3}$$

where n denotes the number of output nodes. After the training process is successfully finished, a ratio template computed for every patch in intensity feature image is used for an input of the AAMLP, and we compute the correlation value between the input and the corresponding output of the AAMLP. If the degree of correlation is above a threshold, we regard the patch contains a face.

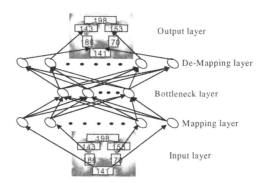

Fig. 5. The architecture of the 4-layer AAMLP model that mimics the top-down mechanism in the IT area. Transfer functions for mapping and de-mapping layers are f(x) = 1/(1+exp(-x)), and those for bottleneck and output layers are linear

The proposed ratio template model is based on the Sinha's method [16]. Sinha used the intensity information of an image to obtain the ratio between the sub-regions in the template. But, the ratio template method using only the intensity information shows less robust to the variation of the illumination. Such a model can detect a non-face area as a face area.

However, in our proposed model, the ratio template is not fixed, but it can be trained as time goes. Through this training mechanism, we can be more robust against the variation of illumination and also might partly solve the rotation or size problem for face detection.

Using the proposed bottom-up saliency map model, the face area is frequently selected as a salient region. Thus, the bottom-up saliency map not only gives selective attention area information but also can give primitive information of face area. Thus, if we consider the bottom-up saliency map information together with the ratio template matching result, we can enhance the performance to localize faces.

As shown in Fig. 3, according to the bottom up saliency map information, the ratio template matching model conducts the matching process for a specific region only in the selected attention area but in all over input image. And, the face ratio template matching model generates face ratio template matching map using the AAMLP model. Finally, the result of the face ratio template matching is integrated with the bottom-up saliency map so that we can make a confidence for the face area and generate the face conspicuity map that represents the face localized information.

3.2 Face Color Perception Model

There have been many proposed model for face detection using face color information [2, 17, 18]. Even though the color information is also important feature for face detection, a face detection model using only the color information has not shown good performance because the color information is also very sensitive to the variations of environment such as illumination and shadows. In order to compensate that problem of using the color information, we also use the bottom-up saliency map model together with face color information. Fig. 3 shows the architecture of the face color perception model hybridized with the bottom-up saliency map model.

According to the biological mechanism, the color opponent feature such as R+G-, B+Y-, B-W+ are used as input of the proposed face color perception model. The proposed model used intensity information (I) for B-W+ color opponent feature (BW). For face detection based on face color perception model, we get the representative face color vector such as (R+G-$_{ref}$, B+Y-$_{ref}$, B-W+$_{ref}$), which is computed by averaging the color values of all the pixels obtained from the face area in the natural color image. The face area is indicated through human interaction. The face area is detected by simple color comparison between the averaged input color vector (R+G-$_p$, B+Y-$_p$, B-W+$_p$) for each patch and the representative color vectors. Then, the result of the face color region detection is integrated with the bottom-up saliency map so that we can make a confidence for the face area. In our bottom-up saliency map model, the face area is frequently perceived as the salient area which shows the coincidence with the human perception for the natural visual scene.

4   Experimental Results and Discussion

Figs. 6 (a) and (b) show the face detection results using only face ratio template matching method like Sinha's model. Figs. 6 (c) and (d) show the face detection results of our proposed model using both the bottom-up saliency map and the face ratio template matching method. As shown in Fig. 6, our proposed model, using both the bottom-up saliency map model and the face ratio template matching method with training mechanism, gives better face detection result. Moreover, our proposed model can reduce tremendously the processing time since the model is searching only the attention area decided by bottom-up selective attention model. Fig. 7 shows the face detection results for another image. Even though the proposed method gives enhanced results for face detection, there exists false detection when non-face area satisfies all the ratios and it is more salient than face area, which means it is selected as an attention area. In such a case, our proposed model does not work properly. In order to localize a face area more accurately, we need to consider an additional feature that sufficiently reflect the characteristics of faces. We considered the face color information which also plays an important role to perceive a human face in natural scene. Fig. 8 shows the result of the proposed face color perception model. As shown in Figs. 8 (c) and (d), however, the face color perception model can not give good performance when a selected area has a lot of similar color with the face color. Using the bottom-up saliency map model together with the face color perception model, we can enhance the face detection performance, but it is not also enough to localize the face area in complex natural scene. Thus, we need to integrate the ratio template matching method with the face color perception model as well as the bottom-up saliency map model as our human does. Fig. 9 shows the results of the proposed integration model. As shown in Fig. 9, the proposed integration model shows a good performance to localize the faces in the natural scene compared with the results in Fig.6-8. Fig. 10 shows an example that our proposed model works properly even when including a naked human with an exposed body and different light condition with training data.

(a)   (b)   (c)   (d)

Fig. 6. The face detection results; (a) and (b) using only face ratio template matching method: (a) the summation result of the intensity information and the result of face ratio template matching (white blocks), (b) face detection result by a threshold of (a); (c) and (d) using the proposed model: (c) the integration of the saliency map and the result of face ratio template matching (white blocks) result, (d) face detection result by a threshold of (c)

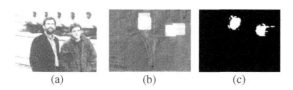

(a)   (b)   (c)

Fig. 7. The face detection result; (a) input image, (b) face conspicuity map using ration template matching method, (c) face detection result by a threshold of (b)

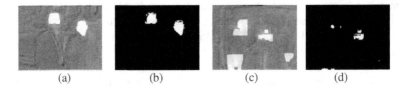

Fig. 8. The experimental results of the proposed face color perception model; (a) and (b) for color input image shown in Fig. 7 (a): (a) face color perception map through the summation of the bottom-up saliency map and the face color detection results (white areas), (b) face detection result by a threshold of (a); (c) and (d) for color input image shown in Fig. 2 (a): (c) and (d) are the same kind of results with (a) and (b), respectively

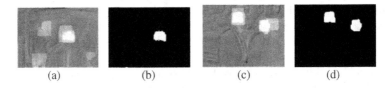

Fig. 9. The experimental results of the proposed face detection model combining the face ratio template matching model and the face color perception model; (a) face conspicuity map for Fig. 2 (a) input image, (b) face detection result by a threshold of (a), (c) face conspicuity map for Fig. 8 (a) input image, (d) face detection result by a threshold of (c)

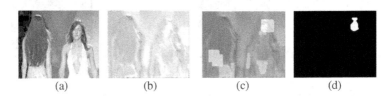

Fig. 10. The experimental results for exposed skin image; (a) input image, (b) saliency map, (c) face conspicuity map, (d) face detection result by a threshold of (c)

5   Conclusion

We proposed a new face detection model based on the bottom up saliency map model as well as the top down face color perception model and the face ratio template matching model with training mechanism. Computer experimental results showed that the proposed face detection method gives a robust performance to detect faces in natural color scenes.

As a further work, we need to consider a size and rotation invariant problem and also more human like binding process for object perception from each specific perceived feature. Also, we are subjected to compare with the performance of others' using complex and comparable benchmarking database. Fundamentally, we are considering the face perception model which can extract automatically the characteristics of a face from a lot of natural scenes and evolve incrementally as time goes.

Acknowledgement

This work was supported by Korea Research Foundation Grant. (KRF-2002-D120303-E00047).

References

1. Itti, L., Koch, C., Niebur, E.: A model of saliency-based visual attention for rapid scene analysis. IEEE Trans. Patt. Anal. Mach. Intell. Vol. 20. 11 (1998) 1254-1259
2. Yang, M., Kriegman, D. J., Ahuja, N.: Detecting faces in images: a survey. IEEE Trans. Patt. Anal. Mach. Intell. vol. 24 no. 1 (2002) 34-58
3. Walther, D., Itti, L., Riesenhuber, M., Poggio, T., Koch, C.: Attentional selection for object recognition – a gentle way. BMCV 2002, Lecture Notes in Computer Science, Vol. 2525, Springer-Verlag, Heidelberg (2002) 472-479
4. Serre, T., Riesenhuber, M., Louie, J., Poggio, T.: On the role of object-specific features for real world object recognition in biological vision. BMCV 2002, Lecture Notes in Computer Science, Vol. 2525, Springer-Verlag, Heidelberg (2002) 387-397
5. Barlow, H.B., Tolhust, D.J.: Why do you have edge detectors? Optical society of America Technical Digest. 23 (1992) 172
6. Bell, A.J., Sejnowski, T.J.: The independent components of natural scenes are edge filters. Vision Research. 37 (1997) 3327-3338
7. Park, S. J., An, K. H., Lee, M.: Saliency map model with adaptive masking based on independent component analysis. Neurocomputing. Vol. 49. (2002) 417-422
8. Guyton, A. C., Textbook of medical physiology, 8th ed., W.B. Saunders company, USA, (1991)
9. Majani, E., Erlanson, R., Abu-Mostafa, Y.: The eye, Academic, New York, (1984)
10. Kuffler, S.W., Nicholls, J.G., Martin, J.G.: From Neuron to Brain. Sinauer Associates, Sunderland, U.K (1984)
11. Bruce Goldstein E.: Sensation and Perception. 4th edn. An international Thomson publishing company, USA (1995)
12. Lee, T.W.: Independent Component Analysis-theory and application. Kluwer academic publisher, USA (1998)
13. Ratnaparkhi, A.: Maximum entropy models for natural language ambiguity resolution, Ph.D thesis from Computer and Information Science in the University of Pennsylvania (1998)
14. Bednar, J.A., Miikkulainen, R.: Self-organization of innate face preferences: Could genetics be expressed through learning? in proceeding of 17th National Conference on Artificial Intelligence, (2000) 117-122
15. Baek, J., Cho, S.: Time jump in?: long rising pattern detection in KOSPI200 future using an auto-associative neural network. in proceeding of 8th International Conference on Neural Information Processing, Shanghai, China, Nov. 14-18, (2001) 160-165
16. Sinha, P.: Qualitative representations for recognition. BMCV 2002, Lecture Notes in Computer Science, Vol. 2525, Springer-Verlag, Heidelberg (2002) 453-461
17. McKennan, S., Gong, S., Raja, Y.: Modeling facial color and identity with Gaussian mixtures. Patten Recognition. vol. 31. no. 12. (1998) 1883-1892
18. Forsyth, M.: A novel approach to color constancy. International Journal of Computer Vision, vol. 5. no.1. (1990) 5-36

Accumulative Computation Method for Motion Features Extraction in Active Selective Visual Attention

Antonio Fernández-Caballero[1], María T. López[1], Miguel A. Fernández[1], José Mira[2], Ana E. Delgado[2], and José M. López-Valles[3]

[1] Universidad de Castilla-La Mancha, E.P.S.A., 02071 - Albacete, Spain
caballer@info-ab.uclm.es
[2] Universidad Nacional de Educación a Distancia,
E.T.S.I. Informática, 28040 - Madrid, Spain
jmira@dia.uned.es
[3] Universidad de Castilla-La Mancha, E.U.P.C., 13071 - Cuenca, Spain
JoseMaria.Lopez@uclm.es

Abstract. A new method for active visual attention is briefly introduced in this paper. The method extracts motion and shape features from indefinite image sequences, and integrates these features to segment the input scene. The aim of this paper is to highlight the importance of the accumulative computation method for motion features extraction in the active selective visual attention model proposed. We calculate motion presence and velocity at each pixel of the input image by means of accumulative computation. The paper shows an example of how to use motion features to enhance scene segmentation in this active visual attention method.

1 Introduction

Findings in psychology and brain imaging have increasingly suggested that it is better to view visual attention not as a unitary faculty of the mind but as a complex organ system sub-served by multiple interacting neuronal networks in the brain [1]. At least three such attentional networks, for alerting, orienting, and executive control have been identified. The images are built habitually as from the entries of parallel ways that process distinct features: motion, solidity, shape, colour, location [2]. Vecera [3] introduced a model to obtain objects separated from the background in static images by combing bottom-up (scene-based) and top-down (task-based) processes. The bottom-up process gets the borders to form the objects, whereas the top-down process uses known shapes stored in a database to be compared to the shapes previously obtained in the bottom-up process. One of the most influential theories about the relation between attention and vision is the Feature Integration Theory [4]. They hypothesized that simple features were represented in parallel across the field, but that their conjunctions could only be recognized after attention had been focused on particular locations. Recognition occurs when the more salient features of the distinct feature maps of features are integrated.

The first neurally plausible architecture of selective visual attention was proposed by Koch and Ullman [5], and is closely related to the Feature Integration Theory.

L. Paletta et al. (Eds.): WAPCV 2004, LNCS 3368, pp. 206–215, 2005.

A visual attention system inspired by the behaviour and the neural architecture of the early primate visual system is presented in [6]. Multiscale image features are combined into a single saliency map. The model of Guided-Search (GS) [7] uses the idea of saliency map to realize the search in scenes. GS assumes a two-stage model of visual selection. The first, pre-attentive stage of processing has great spatial parallelism and realizes the computation of the visual simple features. The second stage is spatially serial and it enables more complex visual representations to be computed, involving combinations of features.

Recently, a neural network (connectionist) model called the Selective Attention for Identification Model (SAIM) has been introduced [8]. The function of the suggested attention mechanism is to allow translation-invariant shape-based object recognition. Also a system of interconnected modules consisting of populations of neurons for modelling the underlying mechanisms involved in selective visual attention is proposed [9]. The dynamics of the system can be interpreted as a mechanism for routing information from the sensory input. A very recent model of attention for active vision has been introduced by Backer and Mertshing [10]. In this model there are two selection phases. Previous to the first selection a saliency map is obtained as the result of integrating the different features extracted. Concretely the features extracted are symmetry, eccentricity, colour contrast, and depth. The first selection stage selects a small number of items according to their saliency integrated over space and time. These items correspond to areas of maximum saliency and are obtained by means of active neural fields. The second selection phase has top-down influences and depends on the system's aim. Some implemented systems based on selective attention have up to date covered up several of the following categories: recognition (e.g. [11]), teleconferencing [12], tracking of multiple objects (e.g. [13]), and mobile robot navigation (e.g. [14]).

In this paper, we briefly describe our approach to selective visual attention [15]. But our intention is to highlight the benefits of using accumulative computation as a method for motion features extraction, as one of the most important contributions to general feature extraction step.

2   Selective Visual Attention Model

The layout of the Selective Visual Attention model developed in our research team is depicted in Figure 1(a). Next a brief description of all tasks involved in our model is offered. The aim of task Attention Construction is to select zones (blobs) of those objects (figures) where attention is to be focused. Notice that after Attention Construction complete figures will not be classified, but all blobs configuring the figures will have been labelled. Blob has to be understood as a homogeneous zone of connected pixels. Therefore, blobs are constructed from image pixels that fulfil a series of predefined requisites (interest points). Task Motion Features Extraction is justified by the need to acquire active features of the image pixels. Concretely, features extracted are "motion presence" and "velocity". Now, task Form Features Extraction computes the values of various shape properties of the objects to be selected. The input to this task is stored as blobs in the Working Memory and as figures in the Attention Focus. Features extracted for the blobs are the size, width and height. As figures stored in the Attention Focus are

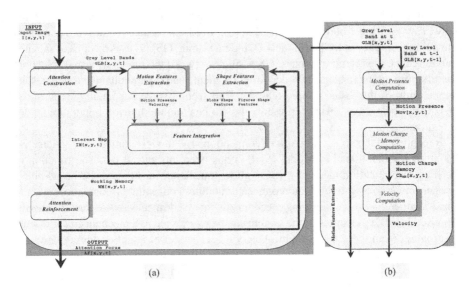

Fig. 1. (a) Selective Visual Attention architecture. (b) Layout of Motion Features Extraction

approximations to complete objects, the features extracted are the same ones than for the blobs, plus features width-height ratio and compactness. The output of task Features Integration is the Interest Map, produced from an integration of motion and form features. Task Attention Reinforcement is dedicated to the final construction of figures and the persistence of attention on some figures (or objects) of interest in the image sequence.

3   Accumulative Computation

Accumulative computation has now been largely applied to moving objects detection, classification and tracking in indefinite sequences of images (e.g. [16], [17], [18], [19]). The more general modality of accumulative computation is the charge/discharge mode, which may be described by means of the following generic formula:

$$Ch[x, y, t] = \begin{cases} \min(Ch[x, y, t - \Delta t] + C, Ch_{max}), & \text{if } "property P[x, y, t]" \\ \max(Ch[x, y, t - \Delta t] - D, Ch_{min}), & \text{otherwise} \end{cases} \quad (1)$$

The temporal accumulation of the persistency of the binary property $P[x, y, t]$ measured at each time instant t at each pixel $[x, y]$ of the data field is calculated. Generally, if the *property* is fulfilled at pixel $[x, y]$, the charge value at that pixel $Ch[x, y, t]$ goes incrementing by increment charge value C up to reaching Ch_{max}, whilst, if *property P* is not fulfilled, the charge value $Ch[x, y, t]$ goes decrementing by decrement charge value D down to Ch_{min}. All pixels of the data field have charge values between the minimum charge, Ch_{min}, and the maximum charge, Ch_{max}. Obviously, values C, D, Ch_{min} and

Ch_{max} are configurable depending on the different kinds of applications, giving raise to all different operating modes of the accumulative computation. Values of parameters C, D, Ch_{max} and Ch_{min} have to be fixed according to the applications characteristics. Concretely, values Ch_{max} and Ch_{min} have to be chosen by taking into account that charge values will always be between them. The value of C defines the charge increment interval between time instants $t-1$ and t. Greater values of C allow arriving in a quicker way to saturation. On the other hand, D defines the charge decrement interval between time instants $t-1$ and t. Thus, notice that the charge stores motion information as a quantified value, which may be used for several classification purposes. In [20] the architecture of the accumulative computation module is shown. Some of the operating modes may be appreciated there, demonstrating their versatility and their computational power.

4   Motion Features Extraction by Accumulative Computation

As told before, the main objective of this paper is to highlight the importance of the accumulative computation method for motion features extraction in the active selective visual attention model proposed. The aim of task Motion Feature Extraction is to calculate the active (motion) features of the image pixels, that is to say, in our case, the presence of motion and the velocity. Due to our experience (e.g. [21]) we know some methods to get that information.

Firstly, in order to diminish the effects of noise due to the changes in illumination in motion detection, variation in grey level bands at each image pixel is treated. We work with 256 grey level input images and transform them to a lower number of levels n. In concrete, good results use to be obtained with 8 levels. These 8 level images are called images segmented into 8 grey level bands and are stored in the Grey Level Bands Map [16], [18], as stated in Equation 2:

$$GLB[x,y,t] = \frac{GL[x,y,t] \cdot n}{GL_{max} - GL_{min} + 1} + 1 \qquad (2)$$

where $GLB[x,y,t]$ is the grey level band of pixel $[x,y]$ at t, GL stands for grey level and n is the total number of grey level bands defined.

In Figure 1(b) you may observe the layout of task Motion Features Extraction. The values computed are Motion Presence, Motion Charge Memory and Velocity. Motion Charge Memory is obtained by means of accumulative computation on the negation of property Motion Presence. Velocity is computed from values stored in Motion Charge Memory. By Velocity we mean the module and angle of vector velocity.

4.1   Motion Presence Computation

The first motion feature calculated is Motion Presence, $Mov[x,y,t]$, which is easily obtained as a variation in grey level band between two consecutive time instants t and $t-1$:

$$Mov[x,y,t] = \begin{cases} 0, \text{ if } GLB[x,y,t] = GLB[x,y,t-1] \\ 1, \text{ if } GLB[x,y,t] \neq GLB[x,y,t-1] \end{cases} \qquad (3)$$

4.2   Motion Charge Memory Computation

As we already stated before, Motion Charge Memory is calculated by means of accumulative computation on the negative of property Motion Presence. The accumulative
computation operation mode used in this case is the LSR (length-speed ratio) mode [22].
The property measured in this case is equivalent to "no motion" at pixel of co-ordinates
$[x, y]$ at instant t.

In this mode C_{MM} (formerly C in Equation 1 is now the charge increment value
on Motion Charge Memory. Notice that D_{MM}, (formerly D) the decrement charge
value does not appear explicitly, as we consider that $D_{MM} = Ch_{max}$. The idea behind
the LSR is that if there is no motion on pixel $[x, y]$, charge value $Ch_{MM}[x, y, t]$ goes
incrementing up to Ch_{max}, and if there exists motion, there is a complete discharge
(the charge value is given value Ch_{min}). Thus, charge value $Ch_{MM}[x, y, t]$ represents a
measure of time elapsed since the last significant variation in brightness on image pixel
$[x, y]$.

$$Ch_{MM}[x, y, t] = \begin{cases} Ch_{min}, & \text{if } Mov[x, y, t] = 1 \\ \min(Ch_{MM}[x, y, t-1] + C_{MM}, \\ Ch_{max}), & \text{if } Mov[x, y, t] = 0 \end{cases} \quad (4)$$

Equation 4 shows how charge at pixel $[x, y]$ gradually increases through time (frame
to frame) in a quantity C_{MM} (charge constant due to motion) up to a maximum charge
or saturation Ch_{max}, while motion is not detected. At the opposite, charge falls down
to a minimum of charge Ch_{min}, when motion is detected at pixel $[x, y]$.

4.3   Velocity Computation

Calculation of velocity is performed starting from the values stored in the Motion Charge
Memory, as explained in Table 1. It is important to highlight that velocity obtained from

Table 1. Description of values stored in Motion Charge Memory

Value in Motion Charge Memory	Explanation
$C_{MM}[x, y, t] = Ch_{min}$	Motion is detected at pixel $[x, y]$ in t. Value in memory is the minimum charge value.
$C_{MM}[x, y, t] = Ch_{min} + k \cdot C_{MM} < Ch_{max}$	No motion is detected at pixel $[x, y]$ in t. Motion was detected for the last time in $t-k \cdot \Delta t$. After k charge increments the maximum charge has not yet been reached.
$C_{MM}[x, y, t] = Ch_{max}$	No motion is detected at pixel $[x, y]$ in t. We do not know when motion was detected for the last time. Value in memory is the maximum charge value.

Motion Charge Memory is not the velocity of an object point that occupies pixel $[x, y]$
in time t, but rather the velocity of an object point that caused motion presence detection
when it passed over pixel $[x, y]$ a number $k = \frac{C_{MM}[x,y,t] - Ch_{min}}{C_{MM}}$ time units ago. Thus,
notice that Motion Charge Memory shows the same value for all those pixels where

a simultaneous motion occurred at a given time. Now, in order to perform Velocity Computation we calculate the velocity in x-axis, v_x, as well as in y-axis, v_y. Once values v_x and v_y, have been obtained, the module and the angle of vector velocity are gotten. Firstly, to calculate velocity in x-axis, charge value in $[x, y]$, where an object is currently passing, is compared to charge value in another co-ordinate of the same row $[x + l, y]$, where the same object is passing. In the best case, that is to say, when both values are different from Ch_{max}, the time elapsed since motion was lastly detected in instant $t - k_{[x,y]} \cdot \Delta t$ at $[x, y]$ up to the time when motion was detected in instant $t - k_{[x+l,y]} \cdot \Delta t$ in $[x + l, y]$ may be calculated as:

$$
\begin{aligned}
Ch_{MM}[x, y, t] - Ch_{MM}[x + l, y, t] &= \\
= (Ch_{min} + k_{[x,y]} \cdot C_{MM}) - (Ch_{min} + k_{[x+l,y]} \cdot C_{MM}) &= \\
= (k_{[x,y]} - k_{[x+l,y]}) \cdot C_{MM}
\end{aligned}
\tag{5}
$$

This computation can obviously not be performed if any of both values are Ch_{max}, as we do not know how many time intervals have elapsed since last movement. Hence, for valid charge values, we have:

$$
\Delta t = \frac{(k_{[x,y]} - k_{[x+l,y]}) \cdot C_{MM}}{C_{MM}} = k_{[x,y]} - k_{[x+l,y]}
\tag{6}
$$

From Equation 5 and Equation 6:

$$
\Delta t = \frac{Ch_{MM}[x, y, t] - Ch_{MM}[x + l, y, t]}{C_{MM}}
\tag{7}
$$

And, as $v_x[x, y, t] = \frac{\delta x}{\delta t} = \frac{l}{\Delta t}$, finally:

$$
v_x[x, y, t] = \frac{C_{MM} \cdot l}{Ch_{MM}[x, y, t] - Ch_{MM}[x + l, y, t]}
\tag{8}
$$

The same way, velocity in y-axis is calculated from the values stored in the Motion Charge Memory, as:

$$
v_y[x, y, t] = \frac{C_{MM} \cdot l}{Ch_{MM}[x, y, t] - Ch_{MM}[x, y + l, t]}
\tag{9}
$$

Now, it is the turn to calculate the module $|\overrightarrow{v}[x, y, t]|$ and the angle $\beta[x, y, t]$ of the velocity.

$$
\beta[x, y, t] = \arctan \frac{v_y[x, y, t]}{v_x[x, y, t]}
\tag{10}
$$

$$
|\overrightarrow{v}[x, y, t]| = (v_x[x, y, t]^2 + v_y[x, y, t]^2)^{0.5}
\tag{11}
$$

5   Data and Results

In order to evaluate the performance of our active visual attention method, and particularly in relation to the motion features described, we have tested the algorithms on

Table 2. Blob shape features and values

Feature	Value (number of pixels)
Spot maximum size	6000
Spot maximum width	85
Spot maximum height	65

Table 3. Figures shape features and values

Feature	Value (pixels)	Value (ratio)
Object size range	400 - 6000	
Object width range	20 - 85	
Object height range	20 - 65	
Object width-height ratio range		0.05 - 2.50
Object compactness range		0.40 - 1.00

the famous Hamburg Taxi motion sequence from the University of Hamburg, usually accepted as an excellent benchmark in optic flow algorithms implementations.

The sequence may be downloaded via ftp://ftp.csd.uwo.ca/pub/vision/, and contains 20 190x256 pixel image frames. Notice that our algorithms only segment moving objects. The sequence contains a movement of four objects: a pedestrian near to the upper left corner and the three cars.

Our intention is to focus only on cars. Thus, we have to parameterize the system in order to capture attention on elements with a series of shape features. These shape features are described in Tables 2 and 3, and are thought to capture all moving cars in the scene. Table 1 shows the parameters used (as well as their values) to get the blobs in the Working Memory. Similarly, in Table 2 we show the parameters and values for the figures in the Attention Focus.

Firstly, results are shown in Figure 2 (upper images) when no predefined velocity is given to the system. In this figure you may appreciate some images of the sequence of selective attention on moving cars. In (a) an input image of the Hamburg Taxi sequence is shown, namely at time instant $t = 9$. In (b) we show in white color the pixels where motion has been detected. Remember that this is equivalent to the result of calculating the presence of motion in the example. Notice that, in the output of this task, a pixel drawn in white color means that there has been variation in the grey level band of the pixel in instant t with respect to the previous instant $t-1$. There are pixels belonging to the desired objects, as well as to other parts of the image due to some variations in illumination in the scene. In (c) see the contents of the Attention Focus. In this figure, pixels drawn in white color on black background represent image elements where attention has been captured and reinforced through time.

In this example we may appreciate that the attention focus really corresponds to moving cars. But, although all moving cars are initially detected - through motion presence feature-, only two of the three cars in movement are segmented. This is due to the fact that the segmentation in grey level bands (as explained in Motion Features Extraction task) unites the moving car to a tree. This union affects our algorithms in a negative way, as the so formed object does not fit into the shape features given in Tables 2 and 3.

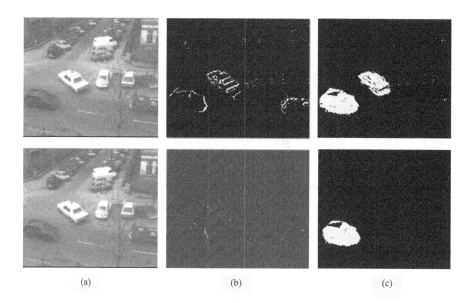

Fig. 2. Sequence of selective attention on moving cars (upper row), and on cars moving to the right (lower row). From top to bottom: (a) Input image. (b) Motion Presence (c) Attention Focus

This example is very helpful to highlight some pros and contras of our method described. Firstly, it is able to discriminate moving objects in an indefinite sequence into different classes of objects. This has been shown by the elimination of the pedestrian in the scene through shape features parameterization. But, clearly, some problems related to temporal overlaps affect our method. Now, consider the lower images at Figure 2, where the attention focus selection has been changed to incorporate velocity parameters. In this case, we are interested in using more motion features to enhance segmentation. Our intention is now to obtain cars that move to the right. This has been accomplished by looking for an angle in vector velocity in the range -22.5° to +22.5°, that is to say:

$$-22.5 \leq \beta[x,y,t] = \arctan \frac{v_y[x,y,t]}{v_x[x,y,t]} \leq +22.5$$

In the results offered in Figure 2 (second row) you may observe that white pixels in (b) image have greatly decreased respect to the results in Figure 2 (first row). That is because only pixels moving with a given velocity angle are filtered. This example shows the importance of motion features to enhance the segmentation in our active visual attention system whilst shape features are maintained constant.

6   Conclusions

A model of dynamic visual attention capable of segmenting objects in a real scene has been briefly described in this paper. The model enables focusing the attention at each moment at shapes that possess certain features and eliminating those that are of no interest. The features used are related to motion and shape of the elements present in the grey level images dynamic scene. The model may be used to observe real environments indefinitely in time.

The principal aim of this paper has been to highlight the importance of the accumulative computation method for motion features extraction in the dynamic selective visual attention model proposed. This is true, because we calculate motion presence and velocity at each pixel of the input image by means of accumulative computation.

Apart from this, our paper highlights the importance of motion features - motion presence and velocity - to enhance the segmentation and classification of objects in real scenes. An example has been offered where, by incrementing the number of motion features, whilst maintaining the shape features constant, the attention focus is changed to the user's interest.

Acknowledgements

This work is supported in part by the Spanish CICYT TIN2004-07661-C01-01 and TIN2004-07661-C02-02 grants.

References

1. Posner, M.I., Raichle, M.E.: Images of Mind. Scientific American Library, NY (1994)
2. Desimone, R., Ungerleider, L.G.: Neural mechanisms of visual perception in monkeys. In: Handbook of Neuropsychology, Elsevier (1989) 267–299
3. Vecera, S.P.: Toward a biased competition account of object-based segregation and attention. In: Brain and Mind, Kluwer Academic Publishers (2000) 353–384
4. Treisman, A.M., Gelade, G.: A feature-integration theory of attention. Cognitive Psychology 12 (1980) 97–136
5. Koch, C., Ullman, S.: Shifts in selective visual attention: towards the underlying neural circuitry. Human Neurobiology 4 (1985) 219–227
6. Itti, L., Koch, C., Niebur, E.: A model of saliency-based visual attention for rapid scene analysis. IEEE Transactions on Pattern Analysis and Machine Intelligence 20 (1998) 1254–1259
7. Wolfe, J.M.: Guided Search 2.0. A revised model of visual search. Psychonomic Bulletin & Review 1 (1994) 202–238
8. Heinke, D., Humphreys, G.W., diVirgilo, G.: Modeling visual search experiments: Selective Attention for Identification Model (SAIM). Neurocomputing 44 (2002) 817–822
9. Deco, G., Zihl, J.: Top-down selective visual attention: A neurodynamical approach. Visual Cognition 8:1 (2001) 119–140
10. Backer, G., Mertsching, B.: Two selection stages provide efficient object-based attentional control for active vision. Proceedings of the International Workshop on Attention and Performance in Computer Vision (2003) 9–16
11. Paletta, L., Pinz, A.: Active object recognition by view integration and reinforcement learning. Robotics and Autonomous Systems 31:1-2 (2000) 71–86
12. Herpers, R., Derpanis, K., MacLean, W.J., Verghese, G., Jenkin, M., Milios, E., Jepson, A., Tsotsos, J.K.: SAVI: An actively controlled teleconferencing system. Image and Vision Computing 19 (2001) 793–804

13. Wada, T., Matsuyama, T.: Multiobject behavior recognition by event driven selective attention method. IEEE Transactions on Pattern Analysis and Machine Intelligence **22**:8 (2000) 873–887

14. Ye, Y., Tsotsos, J.K.: Sensor planning for 3D object search. Computer Vision and Image Understanding **73**:2 (1999) 145–168

15. López, M.T., Fernández, M.A., Fernández-Caballero, A., Delgado, A.E.: Neurally inspired mechanisms for the active visual attention map generation task. Computational Methods in Modeling Computation, Springer-Verlag (2003) 694–701

16. Fernández-Caballero, A., Mira, J., Fernández, M.A., López, M.T.: Segmentation from motion of non-rigid objects by neuronal lateral interaction. Pattern Recognition Letters **22**:14 (2001) 1517–1524

17. Fernández-Caballero, A., Mira, J., Delgado, A.E., Fernández, M.A.: Lateral interaction in accumulative computation: A model for motion detection. Neurocomputing **50** (2003) 341–364

18. Fernández-Caballero, A., Fernández, M.A., Mira, J., Delgado, A.E.: Spatio-temporal shape building from image sequences using lateral interaction in accumulative computation. Pattern Recognition **36**:5 (2003) 1131–1142

19. Fernández-Caballero, A., Mira, J., Fernández, M.A., Delgado, A.E.: On motion detection through a multi-layer neural network architecture. Neural Networks **16**:2 (2003) 205–222

20. Mira, J., Fernández, M.A., López, M.T., Delgado, A.E., Fernández-Caballero, A.: A model of neural inspiration for local accumulative computation. 9th International Conference on Computer Aided Systems Theory, Springer-Verlag (2003) 427–435

21. Fernández, M.A., Mira, J.: Permanence memory: A system for real time motion analysis in image sequences. IAPR Workshop on Machine Vision Applications (1992) 249–252

22. Fernández, M.A., Fernández-Caballero, A., López, M.T., Mira, J.: Length-speed ratio (LSR) as a characteristic for moving elements real-time classification. Real-Time Imaging **9** (2003) 49–59

Fast Detection of Frequent Change in Focus of Human Attention

Nan Hu[1,2], Weimin Huang[1], and Surendra Ranganath[2]

[1] Institute for Infocomm Research (I²R),
21 Heng Mui Keng Terrace, Singapore 119613
{nhu, wmhuang}@i2r.a-star.edu.sg
[2] Department of Electrical and Computer Engineering,
National University of Singapore,
4 Engineering Drive 3, Singapore 117576
elesr@nus.edu.sg

Abstract. We present an algorithm to detect the attentive behavior of persons with frequent change in focus of attention (FCFA) from a static video camera. This behavior can be easily perceived by people as temporal changes of human head pose. Here, we propose to use features extracted by analyzing a similarity matrix of head pose by using a self-similarity measure of the head image sequence. Further, we present a fast algorithm which uses an image vector sequence represented in the principal components subspace instead of the original image sequence to measure the self-similarity. An important feature of the behavior of FCFA is its cyclic pattern where the head pose repeats its position from time to time. A frequency analysis scheme is proposed to find the dynamic characteristics of persons with frequent change of attention or focused attention. A nonparametric classifier is used to classify these two kinds of behaviors (FCFA and focused attention). The fast algorithm discussed in this paper yields real-time performance as well as good accuracy.

1 Introduction

Computation for detecting attentive behavior has long been focusing on the task of selecting salient objects or short-term motion in images. Most of the research works use low level attentive vision with local features such as edge, corner, color and motion etc.[1, 2, 3]. In contrast, our work deals with the issue of detecting salient motions from long-term video sequences, i.e. global attentive behavior, which is a new topic in this area.

As we know, human attention can always be directed to salient objects selected by the observer. When salient objects are spatially widely distributed, however, visual search for the objects will cause a frequent change in focus of attention (FCFA). For example, the number of salient objects to a shopper can be extremely large, therefore, in a video sequence, the shopper's attention will change frequently. This kind of attentive motion can be a salient cue for computer recognition of attention.

On the other side, when salient objects are localized, visual search will cause human attention to focus on one spot only, resulting in focused attention. In this paper we propose

L. Paletta et al. (Eds.): WAPCV 2004, LNCS 3368, pp. 216–230, 2005.

to classify these two kinds of attentive behavior. Although FCFA can be easily perceived by humans as temporal changes of head pose which keeps repeating itself in different orientations, it is not easily measured directly from the sequence that demonstrates a cyclic motion pattern. Contrary to FCFA, an ideally focused attention implies that head pose remains unchanged for a relatively long time, i.e., no cyclicity is demonstrated. Our work, therefore, is to mimic human perception of FCFA as a cyclic motion of a head and present an approach for the detection of this cyclic attentive behavior from video sequences.

1.1   Cyclic Motion

The motion of a point $\overline{X}(t)$, at time t, is cyclic if it repeats itself with a time varying period $p(t)$, i.e.,

$$\overline{X}(t + p(t)) = \overline{X}(t) + \overline{T}(t), \tag{1}$$

where $\overline{T}(t)$ is a translation of the point. The period $p(t)$ is the time interval that satisfies (1). If $p(t) = p_0$, i.e., a constant for all t, then the motion is periodic[4]. A periodic motion has a fixed frequency $1/p_0$. However, the frequency of cyclic motion is time varying. Over a period of time, cyclic motion will cover a band of frequencies while periodic motion covers only a single frequency or at most a very narrow band of frequencies.

1.2   Our Approach

Most of the time, the attention of a person can be characterized by his/her head pose[5]. Thus, the underlying change of attention can be inferred by the motion pattern of head pose changes with time. Our approach is to learn the motion pattern by analyzing the motion feature. For FCFA, the human head keeps repeating the poses, which therefore demonstrates cyclic motion as defined above. An obvious measurement for the cyclic pattern is the similarity measure of the frames in the video sequence.

By calculating the self-similarities between any two frames in the video sequence, a similarity matrix can be constructed. As shown later in this paper, a similarity matrix for cyclic motion differs from that of one with smaller motion, which in texture can be distinguished by human eyes easily, such as a video of a person with focused attention.

Since the calculation of the self-similarity using the original video sequence is very time consuming, we further improved the algorithm by using a sequence of image vectors represented in the principal components subspace instead of the original image sequence for the self-similarity measure. This approach saves much computation time and makes the system near real-time with improved classification accuracy, as shown later.

On the similarity matrix we applied a 2-D Discrete Fourier Transform to find the characteristics in the frequency domain. A four dimensional feature vector of normalized Fourier spectrum in the low frequency region is extracted as the feature vector.

Because of the relatively small size of training data, and the unknown distribution of the two classes, we employ a nonparametric classifier, i.e., k-Nearest Neighbor Rule (K-NNR), for the classification of the FCFA and focused attention.

1.3   Related Work

In most cases, focus of attention can be characterized by head orientation[5]. Stiefelhagen used separate neural networks to estimate the pan and tilt of the head pose. He detected a person's attention based on the head orientation.

One of the cues directly related to human attention detection is the head pose. Wu and Toyama [6] built a 3-D ellipsoidal model of the head for each pose using 4 image-based features and determined the pose in an input image by computing the maximum of a *posteriori* pose. Rae and Ritter [7] used three neural networks to do color segmentation, face localization, and head orientation estimation respectively. Zhao and Pingali [8] applied two neural networks to determine pan and tilt angles separately to estimate head orientation. Krüger and Bruns [9] estimated the head orientation using a unique representation based on Gabor Wavelet Network. All these approaches tried to find the orientation of the head in individual images. Our approach of FCFA detection does not require any knowledge of head pose estimation from individual images. Furthermore, in many cases, the head image is so small that it is difficult to estimate head pose.

Another category of research is on cyclic motion detection [10]. However most of the works focused on the periodicity only e.g. [11, 12] using pixel-based methods. Cutler and Davis proposed to use area-based correlation for periodic analysis which shows some advantages [4]. In this paper we will adopt a similar area-based similarity measurement. Different from that in [4], where 1D Fourier Transform and the Short Time Fourier Transform is used to find the peak-frequency or fundamental frequency for the periodic motion, a 2D fundamental Fourier analysis is applied and a learning process is proposed for one cannot find the peak frequency in a non-periodic but cyclic motion.

1.4   Contributions

The main contribution of our work is the introduction of a scheme for the robust analysis of cyclic time-series image data as a whole instead of individual images to detect the behavior of FCFA. Although there were some works presented by other researchers for periodic motion detection, we believe our approach is new to address the cyclic motion problem. Different from the works in head pose detection, our approach requires no information of the exact head pose. Instead, by extracting the global motion pattern from the whole image sequence we detect behaviors. Combined with a simple classifier, the experiments show the robustness of the proposed method for FCFA detection. A fast algorithm is also proposed with improved accuracy for the attentive behavior detection. With the proposed method, a future system could zoom into the scene and extract other detailed information such as face, age, and gender of a person. Obviously it can make a system more computationally efficient by paying more attention on the persons with predefined behaviors(FCFA or focused attention).

2   Detection of FCFA

The algorithm for cyclic motion detection consists of three parts: (1) dimensionality reduction of head images; (2) head pose similarity computation as it evolves in time; (3) frequency analysis and classification. Head tracking is by itself a research area with

Frame Sequence of Head

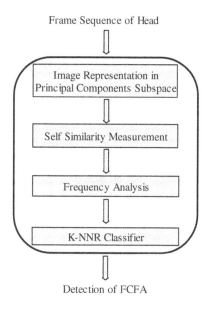

Detection of FCFA

Fig. 1. Overview of our FCFA detection algorithm

several prior works[13, 14]. Here, by manually locating the center of the head of interest in each frame of the video sequence, we decouple head tracking and cyclic motion detection. In this manner, our algorithm can be used with different head tracking algorithms. Figure 1 shows an overview of our algorithm.

In the following sections, video sequences of a person looking around (Watcher), i.e., exhibiting FCFA, and a person standing and talking to others (Talker), i.e., focused attention, will be used to illustrate the technical points.

2.1 Data Description and Preparation

The first set of video sequences is taken by a camera from the overhead corner of a hall with frame rate of 25 frames per second. These sequences are first cropped to a length of forty seconds and then resampled by keeping one out of every 5 frames. Thus, for each person, we get an image sequence of 200 frames. The enlarged head images of Watcher and Talker are shown in Fig 2 and 3, respectively.

Fig. 2. Illustration of extracted heads of a watcher

Fig. 3. Illustration of extracted heads of a talker

The sizes of the head in Fig 2 and Fig 3 are around 29×21 (for Watcher) and 45×31 (for Talker). Since the fixed camera is far away from the object, the head scale in each sequence will be the same. However for different sequences, the head sizes may be different.

2.2   Similarity Matrix

The input data here is a sequence of frames with head centers C_i located. To characterize the cyclicity of the head, we first compute the head H_t's similarity at times t_1 upon t_2. While many image similarity metrics can be used, we used the averaged absolute difference[4, 10], for it needs less computational effort:

$$S_{t_1,t_2} = \frac{1}{N_{B_t}} \sum_{(x,y)\in B_t} |O_{t_1}(x,y) - O_{t_2}(x,y)|, \tag{2}$$

where $O_t(x,y)$ is the image intensity of the pixel (x,y), B_t is the bounding box of head H_t, with the center C at the head center C_{H_t} and N_{B_t} is the total number of pixels in the bounding box B_t. In order to account for head locating errors, the minimal S is found by translating the center of the head H_{t_1} over a small searching square.

$$S'_{t_1,t_2} = \min_{|dx|,|dy|<a} \frac{1}{N_{B_t}} \sum_{(x,y)\in B_t} |O_{t_1}(x+dx, y+dy) - O_{t_2}(x,y)|. \tag{3}$$

Our experiment ($a = 2$ for all sequences) shows that the results are insensitive to the value of a when $a \geq 2$. After we get head H_t's similarity at time t_1 and t_2, we define the similarity matrix of an N-image sequence as

$$R_{H_t} = \left[S'_{t_i,t_j}\right]_{N \times N}, i,j = 1,2,\cdots,N. \tag{4}$$

Figure 4 shows images of the similarity matrix R of Watcher and Talker. The values of the matrix elements have been linearly scaled to the gray-scale intensity range [0,255]. Dark regions show more similarity. Note that the matrix is symmetric along the main diagonal.

Watcher

Talker

Watcher

Talker

Fig. 4. Plot of similarity matrix R **Fig. 5.** Plot of similarity matrix R'

2.3   Dimensionality Reduction and Fast Algorithm

Calculating the absolute difference using original images does work, however, it is time consuming because of the high dimensionality of head images (e.g. the head image for Watcher is a $30 \times 20 = 600$ dimensional vector after normalization). A direct and easy way to save computational time is to use principal component analysis (PCA) to reduce the dimensionality of the images.

Theorem 1. *For any two n-dimensional vectors,* $\mathbf{x} = (x_1, \cdots, x_n)^{\mathrm{T}}$ *and* $\mathbf{y} = (y_1, \cdots, y_n)^{\mathrm{T}}$, *let* $D_{\mathrm{E}}(\mathbf{x}, \mathbf{y})$ *be the Euclidean distance between* $\mathbf{x}$ *and* $\mathbf{y}$ *and* $D_{\mathrm{Abs}}(\mathbf{x}, \mathbf{y})$ *be the absolute distance between* $\mathbf{x}$ *and* $\mathbf{y}$. *Then,*

$$D_{\mathrm{E}}(\mathbf{x}, \mathbf{y}) \leq D_{\mathrm{Abs}}(\mathbf{x}, \mathbf{y}) \leq \sqrt{n} D_{\mathrm{E}}(\mathbf{x}, \mathbf{y}). \tag{5}$$

Corollary 1. *Given any two n-dimensional vectors* $\mathbf{x}$ *and* $\mathbf{y}$, *and let their PCA-transformed m-dimensional vectors be* $\mathbf{x}'$ *and* $\mathbf{y}'$ *respectively. If PCA transform preserves almost all of the energy, the difference between the absolute distance in the original space* $D_{\mathrm{Abs_{Org}}} = D_{\mathrm{Abs}}(\mathbf{x}, \mathbf{y})$ *and that in PCA subspace* $D_{\mathrm{Abs_{PCA}}} = D_{\mathrm{Abs}}(\mathbf{x}', \mathbf{y}')$ *is bounded by* $(D_{\mathrm{E_0}} = D_{\mathrm{E}}(\mathbf{x}, \mathbf{y}))$

$$(1 - \sqrt{m})D_{\mathrm{E_0}} \leq (D_{\mathrm{Abs_{Org}}} - D_{\mathrm{Abs_{PCA}}}) \leq (\sqrt{n} - 1)D_{\mathrm{E_0}} \tag{6}$$

Corollary 1 shows that when $\mathbf{x}$ is near and similar to $\mathbf{y}$, i.e. D_{E} is small, the difference between $D_{\mathrm{Abs_{Org}}}$ and $D_{\mathrm{Abs_{PCA}}}$ is narrowly bounded and from Theorem 1 both $D_{\mathrm{Abs_{Org}}}$ and $D_{\mathrm{Abs_{PCA}}}$ are small too. When $\mathbf{x}$ is far away from and dissimilar to $\mathbf{y}$, i.e. D_{E} is large, from Theorem 1 both $D_{\mathrm{Abs_{Org}}}$ and $D_{\mathrm{Abs_{PCA}}}$ are large. Hence, similar to $D_{\mathrm{Abs_{Org}}}$, $D_{\mathrm{Abs_{PCA}}}$ is also a good way to measure the similarity between images.

In our case, because the different size of head images, we first normalize the image by resizing them to the same size of 30×20. After training, we choose a 9-dimensional PCA subspace, which carries 98% of total energy, to represent the images. The projection matrix from original image space to PCA subspace is $P = P_{9 \times 600}$.

To account for the head center locating error, for the kth head image, we shifted the head center of every head image by ± 1 pixel vertically or horizontally or both, which resulted in 9 possible head images, written as vectors $\mathbf{H}_{k1}, \cdots, \mathbf{H}_{k9}$, each of which is 600-dimensional. It is easy to find out that it covers the same shifting window as that used in Eq (3). Projecting them onto the predefined PCA subspace, we get the 9 vectors

$$\mathbf{h}_{ki} = P \times \mathbf{H}_{ki}, \quad i = 1, \cdots, 9 \tag{7}$$

The similarity between image n and image m is obtained by choosing the minimal pairwise absolute distances in the PCA subspace between the possible head vectors for these two images.

$$S''_{n,m} = \min_{i,j} D_{\mathrm{Abs}}(\mathbf{h}_{ni}, \mathbf{h}_{mj}), \quad i, j = 1, \cdots, 9 . \tag{8}$$

The algorithm is described below:

– For the 1st frame, firstly resize the head size to 30×20. Secondly, shift the head center to get the 600-dimensional head image vectors $\mathbf{H}_{11}, \cdots, \mathbf{H}_{19}$. Thirdly, map them to PCA subspace to get the vectors $\mathbf{h}_{11}, \cdots, \mathbf{h}_{19}$ and store them in memory;

– For the kth ($k = 2, \cdots, 200$) frame, firstly resize the head size to 30×20. Secondly, shift the head center to get the 600-dimensional head image vectors $\mathbf{H}_{k1}, \cdots, \mathbf{H}_{k9}$. Thirdly, map them to PCA subspace to get the vectors $\mathbf{h}_{k1}, \cdots, \mathbf{h}_{k9}$ and store them in memory. Fourthly, calculate the absolute distance S'' between itself and the previous $k - 1$ images;

$$S''_{t,k}, \quad t = 1, \cdots, k - 1 \tag{9}$$

– For the kth image, $k > 200$, we only need to calculate $S''_{t,k}, t = k - 1, \cdots, k - 199$;

– Set $S''_{k,t} = S''_{t,k}$, where $k > t$, and $S''_{k,k} = 0$. Form the similarity matrix

$$R'_H = \left[S''_{i,j} \right], i, j = k - 199, \cdots, k. \tag{10}$$

Figure 5 shows images of the similarity matrix R and R' of Watcher and Talker calculated in the PCA subspace. The values of the matrix elements have been linearly scaled to the gray-scale intensity range [0,255]. Note that as shown similarity matrices R''s are similar to R's in texture except that they are darker than R's.

2.4   Frequency Analysis

For cyclic motion, one of the most important features is the time-frequency information it carries. While many time series analysis methods can be used, we used Fourier analysis for its simplicity and ease of use.

To find the characteristics of the behavior, one way is to apply a 1-D Fourier Transform to the similarity matrix R. Then we average the Fourier spectra of all rows. Figure 6(a) shows the averaged Fourier Spectra of Watcher and Talker. As shown, these spectra for Watcher and Talker are similar. However, if we zoom into the low frequency area, as shown in Figure 6(b), we can see that the spectral values for Talker are larger than those for Watcher.

Since 1-D Fourier Transform can only yield frequency information along rows or columns of the similarity matrix R, and not both, instead of a 1-D Fourier Transform used in [4], a 2-D Discrete Fourier Transform [15] is used here to find the Fourier Spectrum (FS) matrix of the similarity matrix.

To reduce sensitivity to lighting effects, we normalized the Fourier Spectrum matrix FS_R by averaging over the total energy of the similarity matrix R_H,

$$FS'_R = \frac{FS_R}{\sum_{x=1}^{N} \sum_{y=1}^{N} |R_H(x,y)|^2}, \tag{11}$$

where $N = 200$ is the number of images in the sequence.

Central areas of FS'_R and $FS'_{R'}$ matrices of Watcher and Talker are shown in Figures 7 and 8. The values of the elements have been linearly scaled to [0,255]; as the DC component here is much larger than that of any other frequency, we set it the value of

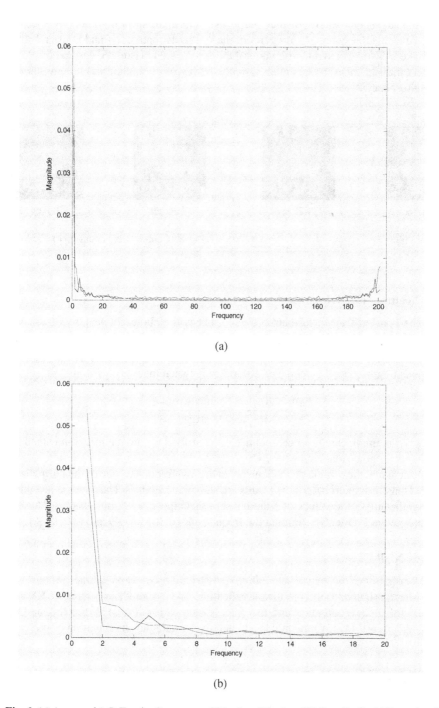

(a)

(b)

Fig. 6. (a) Averaged 1-D Fourier Spectrum of Watcher (Blue) and Talker (Red); (b)Zoom-in of (a) in the low frequency area

the second largest element for display purposes; white areas show high Fourier Spectral values. Note that the symmetric property of the similarity matrix R_H and Fourier Transform make FS' matrix symmetric about the lines $y = \pm x$. Using R' instead of R did not effect the FS' matrix very much visually, but saves computational time as shown later in this paper.

| Walker | Talker | Watcher | Talker |

Fig. 7. Central area zoom-in of FS matrix

Fig. 8. Central area zoom-in of FS' matrix

2.5   Feature Selection

Given the Fourier Transform matrix $FS'_{R'}$, next is the choice of feature vector. We want to use those elements of the Fourier Transform matrix $FS'_{R'}$ that show significant differences between the two classes. Thus, given a FS matrix element e_j, firstly we define a coefficient δ to reflect the degree of difference between the two classes as:

$$\delta_j = \frac{\|\text{mean}(e_j|\omega_1) - \text{mean}(e_j|\omega_2)\|_E}{\text{std}(e_j|\omega_1) + \text{std}(e_j|\omega_2)} \quad (12)$$

where $\|\cdot\|_E$ is Euclidean distance, $\text{mean}(e_j|\omega_i)$, $\text{std}(e_j|\omega_i)$ are the mean and standard deviation of e_j given class ω_i, where $i = 1, 2$.

We calculated the δ value of 16 elements in the low frequency area of the normalized Fourier Transform matrix FS'; the results are shown in Figure 9. The 4 elements which have significantly large values of δ are chosen to compose the feature vector. These 4 elements are the Fourier Spectrum at the frequencies $(0, 0)$, $(0, \frac{2\pi}{N})$ and $(\frac{2\pi}{N}, \pm\frac{2\pi}{N})$.

2.6   K-NNR Classifier

As the distribution of the feature vector is unknown, we employ a nonparametric approach — *k-nearest-neighbor rule* [16] for classification. We assign Class ω_1 for FCFA and Class ω_2 for focused attention and use $k = 3$ (odd to avoid ties). A Leave-One-Out Cross-validation (LOOCV) method is adopted to estimate the classification error.

3   Experimental Results

In this section we present the results of our algorithm on labeled data. In the first experiment, we use a small data set captured from a fixed camera, and cross-validation is used to estimate the classification error. To have a good estimate of the performance, we

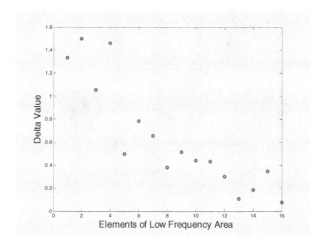

Fig. 9. The δ values of the 16 elements in the low frequency area

conducted a second experiment with more data captured from different cameras and settings to validate the classifier built using all the data in the first experiment. The original head size of the data in both experiments ranges from 25×15 to 63×43.

3.1 Classification and Validation

The labeled training data here include five ω_1 sequences and six ω_2 sequences of different persons. The similarity matrix R and R' for each person are shown in Figure 10 and Figure 11.

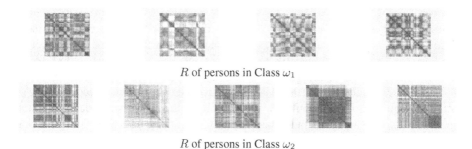

R of persons in Class ω_1

R of persons in Class ω_2

Fig. 10. Similarity matrix R (the original images are omitted here and the R's for another Watcher and Talker are shown in Fig 4)

The results of LOOCV using R showed that none of the ω_1 data in 5 cases was misclassified while one of the ω_2 data in 6 cases was misclassified. When examining the cause of the misclassification (the similarity matrix is the leftmost of Class ω_2 in Figures 9 & 10), we find that the person was listening to others at first and then kept

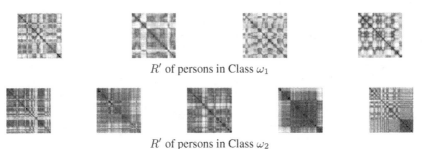

R' of persons in Class ω_1

R' of persons in Class ω_2

Fig. 11. Similarity matrix R' (the original images are omitted here and the R''s for another Watcher and Talker are shown in Fig 5)

Fig. 12. Sampled images of misclassified data in the first experiment using R

changing his attention to other directions (as sampled in Figure 12). Thus, his data is, to some extent, similar to and overlaps with FCFA. As shown in Table 1, however, none is misclassified by R'.

3.2   More Data Validation

To test whether the proposed method generalized well on other data sets, some more video sequences are validated on the classifier which is built with all of the data in Section 3.1. The new sequences include 10 ω_1 sequences and 10 ω_2 sequences with different persons, different head sizes and different camera exposures taken by different cameras.

Using R, the results showed that 2 in 10 ω_1 sequences were misclassified and none of the 10 ω_2 sequences is misclassified by R. Examining the misclassified data, we found that the two ω_1 data are taken under the same illumination and the same exposure which are the lowest among the whole data set. Their faces are dark and almost of the same color as that of hair. Thus, it is reasonable to expect that they would be misclassified.

Using R', however, only one sample was misclassified, yielding an improvement in classification accuracy. One possible reason for the better performance is that mapping the sequences into a subspace reduced the illumination effect while maintain the relative change between frames.

Table 1 summarizes the results of both experiments.

Table 1. Summary of experimental results

		using R		using R'	
		ω_1	ω_2	ω_1	ω_2
First	Class ω_1	4	1	5	0
Experiment	Class ω_2	0	6	0	6
	Accuracy	90.9%		100%	
Second	Class ω_1	10	0	10	0
Experiment	Class ω_2	2	8	1	9
	Accuracy	90%		95%	
Average Accuracy		90.3%		96.8%	

Table 2. Time used to calculate R & R' in Matlab

R	R'
186.3s	73.4s

3.3   Real-Time System

In the system, images are acquired at an interval of 0.2s. The algorithm is implemented using video sequences length of 200 images on a 2.4GHz Pentium IV PC. The most time-consuming step is the calculation of similarity matrix. Compared to this, the time used for FFT and K-NNR is trivial—63ms and 15ms respectively in Matlab. As Table 2 shows, running the algorithm to calculate the whole similarity matrix in Matlab needs 73.4s using R', which is about 2.5 times faster than that using R which needs 186.3s. In a real-time system, upon the arrival of each image, we only need to compute the similarity between itself and the previous 199 images. The computation time is 0.75s in Matlab. It would be in real time if programmed in the C environment.

4   Discussion

For FCFA, the person frequently changes his head pose (this can be achieved by rotating his head or rotating his body or both), which results in the similarity matrix R of the person demonstrating cyclicity. However, for focused attention, the person seldom rotates his head, resulting in the similarity matrix R demonstrating little or no cyclicity. Thus, by 2-D Fourier Transform, after normalizing the total energy, we extract from FS matrix DC component as well as those of three lowest frequencies as features.

Our algorithm is robust to low resolution and varying illumination. The lowest resolution of the head was 25×15 in the experiments. In addition, the R' is noise tolerant since PCA can denoise the raw data. Furthermore, our algorithm is robust to error in head location by searching the minimal S' in a small area that reduces the locating error.

Here, we assume the direction of visual attention is fully characterized by the head pose. However, gaze is also a useful hint to estimate human visual attention. The reason we ignore this kind of detection is that the head image we used in the experiment is relatively small and sometimes the camera cannot even see the eyes, which makes gaze detection very difficult. Besides, in many cases, in order to cover a big area, it is more convenient for people to change the head direction to shift their attention, which motivated the development of the proposed method.

5   Conclusion

Attentive behavior detection is useful for human computer interaction, to know where a person is looking at to further improve the interactivity, or for statistics in education and business, to know if students are focusing on the lecture or inferring whether a product is attractive. To infer this behavior, we have presented a system to detect FCFA. By manually locating the center of the head, we extract a window which includes the complete head. A similarity matrix is computed in the 9-dimensional principal components subspace as the head pose evolves over time. A 2-D frequency analysis is applied on the similarity matrix for feature extraction. Finally, K-NNR is proposed to differentiate FCFA from focused attention.

Future work includes integration of a possible tracking algorithm to locate the head automatically and a more detailed differentiation of head rotations both temporally and spatially, which can be used to detect detailed motion of human such as reading, talking, looking around and so on. Furthermore, this method can also be extended to video summarization and segmentation.

References

[1] Bauer, B., Jolicoeur, P., Cowan, W.B.: Visual search for color targets that are or are not linearly separable from distractors. Vision Research **36** (1996) 1439–1465
[2] Swain, M.J., Ballard, D.H.: Color indexing. Int'l Journal of Computer Vision **7** (1991) 11–32
[3] Tsotsos, J.K., Culhane, S.M., Wai, W.Y.K., Lai, Y., Davis, N., Nuflo, F.: Modeling visual attention via selective tuning. Artificial Intelligence **78** (1995) 507–545
[4] Cutler, R., Davis, L.: Robust real-time periodic motion detection, analysis, and applications. IEEE Trans. Pattern Analysis and Machine Intelligence **22** (2000) 781–796
[5] Stiefelhagen, R.: Tracking focus of attention in meetings. Proc. Fourth IEEE Int'l Conf. Multimodal Interfaces (2002) 273–280
[6] Wu, Y., Toyama, K.: Wide-range person- and illumination-insensitive head orientation estimation. Proc. Fourth Int'l Conf. Automatic Face and Gesture Recognition (2000) 183–188
[7] Rae, R., Ritter, H.: Recognition of human head orientation based on artificial neural networks. IEEE Trans. Neural Networks **9** (1998) 257–265
[8] Zhao, L., Pingali, G., Carlbom, I.: Real-time head orientation estimation using neural networks. Proc. Int'l Conf. Image Processing (2002)
[9] Krüger, V., Bruns, S., Sommer, G.: Efficient head pose estimation with gabor wavelets. Proc. 11th British Machine Vision Conference **1** (2000) 72–81

[10] Seitz, S.M., Dyer, C.R.: View-invariant analysis of cyclic motion. Int'l J. Computer Vision **25** (1997) 1–23

[11] Liu, F., Picard, R.: Finding periodicity in space and time. Proc. Int'l Conf. Computer Vision (1998) 376–383

[12] Polana, R., Nelson, R.: Detection and recognition of periodic, non-rigid motion. Int'l J. Computer Vision **23** (1997) 261–282

[13] Zeng, Z., Ma, S.: Head tracking by active particle filtering. Proc. Fifth IEEE Int'l Conf. Automatic Face and Gesture Recognition (2002) 82–87

[14] Basu, S., Essa, I., Pentland, A.: Motion regularization for model-based head tracking. Proc. IEEE Int'l Conf. Pattern Recognition **3** (1996) 611–616

[15] Gonzalez, R., Woods, R.: Digital Image Processing. Addison-Wesley Publishing Company (1992)

[16] Duda, R., Hart, P., Stork, D.: Pattern Classification. 2nd edn. John Wiley & Sons, Inc. (2000)

A   Appendix

Theorem 1. *For any two n-dimensional vectors, $\mathbf{x} = (x_1, \cdots, x_n)^{\mathrm{T}}$ and $\mathbf{y} = (y_1, \cdots, y_n)^{\mathrm{T}}$, let $D_{\mathrm{E}}(\mathbf{x}, \mathbf{y})$ be the Euclidean distance between $\mathbf{x}$ and $\mathbf{y}$ and $D_{\mathrm{Abs}}(\mathbf{x}, \mathbf{y})$ be the absolute distance between $\mathbf{x}$ and $\mathbf{y}$. Then, we have*

$$D_{\mathrm{E}}(\mathbf{x}, \mathbf{y}) \leq D_{\mathrm{Abs}}(\mathbf{x}, \mathbf{y}) \leq \sqrt{n} D_{\mathrm{E}}(\mathbf{x}, \mathbf{y}).$$

Proof. (of Theorem 1) According to the definition of Euclidean distance and Absolute distance, we have

$$D_{\mathrm{E}}(\mathbf{x}, \mathbf{y}) = \sqrt{\sum_{i=1}^{n} (x_i - y_i)^2}, \tag{1}$$

$$D_{\mathrm{Abs}}(\mathbf{x}, \mathbf{y}) = \sum_{i=1}^{n} |x_i - y_i|$$

$$= \sqrt{(\sum_{i=1}^{n} |x_i - y_i|)^2}, \tag{2}$$

and

$$(\sum_{i=1}^{n} |x_i - y_i|)^2 = \sum_{i=1}^{n} (x_i - y_i)^2 + \sum_{i=1}^{n} \sum_{j=1, j \neq i}^{n} (|(x_i - y_i| \cdot |x_j - y_j|). \tag{3}$$

Since

$$0 \leq \sum_{i=1}^{n} \sum_{j=1, j \neq i}^{n} (|(x_i - y_i| \cdot |x_j - y_j|) \leq (n-1) \cdot \sum_{i=1}^{n} (x_i - y_i)^2, \tag{4}$$

where min appears iff at most one $x_i \neq y_i$, $i = 1 \cdots n$, and max appears iff $|x_i| = |y_i|$, for $\forall i = 1 \cdots n$, we have

$$D_{\mathrm{E}}(\mathbf{x}, \mathbf{y}) = \sqrt{\sum_{i=1}^{n}(x_i - y_i)^2} \leq D_{\mathrm{Abs}}(\mathbf{x}, \mathbf{y}) = \sum_{i=1}^{n}|x_i - y_i| \leq$$

$$\sqrt{n}D_{\mathrm{E}}(\mathbf{x}, \mathbf{y}) = \sqrt{(n-1) \cdot \sum_{i=1}^{n}(x_i - y_i)^2} \tag{5}$$

Corollary 1. *Given any two n-dimensional vectors* $\mathbf{x}$ *and* $\mathbf{y}$, *and let their PCA-transformed m-dimensional vectors be* $\mathbf{x}'$ *and* $\mathbf{y}'$ *respectively. If PCA transform preserves almost all of the energy, the difference between the absolute distance in the original space* $D_{\mathrm{Abs_{Org}}} = D_{\mathrm{Abs}}(\mathbf{x}, \mathbf{y})$ *and that in PCA subspace* $D_{\mathrm{Abs_{PCA}}} = D_{\mathrm{Abs}}(\mathbf{x}', \mathbf{y}')$ *is bounded by* $(D_{\mathrm{E}_0} = D_{\mathrm{E}}(\mathbf{x}, \mathbf{y}))$

$$(1 - \sqrt{m})D_{\mathrm{E}_0} \leq (D_{\mathrm{Abs_{Org}}} - D_{\mathrm{Abs_{PCA}}}) \leq (\sqrt{n} - 1)D_{\mathrm{E}_0}.$$

Proof. (of Corollary 1) By the assumption that PCA transform preserves almost all of the energy, we have

$$D_{\mathrm{E}}(\mathbf{x}', \mathbf{y}') \approx D_{\mathrm{E}}(\mathbf{x}, \mathbf{y}) \overset{\triangle}{=} D_{\mathrm{E}_0}. \tag{6}$$

From Theorem 1, we get

$$D_{\mathrm{E_{Org}}} = D_{\mathrm{E}_0} \leq D_{\mathrm{Abs_{Org}}} \leq \sqrt{n}D_{\mathrm{E_{Org}}} = \sqrt{n}D_{\mathrm{E}_0}, \tag{7}$$

$$D_{\mathrm{E_{PCA}}} \approx D_{\mathrm{E}_0} \leq D_{\mathrm{Abs_{PCA}}} \leq \sqrt{n}D_{\mathrm{E_{PCA}}} \approx \sqrt{n}D_{\mathrm{E}_0}. \tag{8}$$

Subtract (8) from (7), we get

$$(1 - \sqrt{m})D_{\mathrm{E}_0} \leq (D_{\mathrm{Abs_{Org}}} - D_{\mathrm{Abs_{PCA}}}) \leq (\sqrt{n} - 1)D_{\mathrm{E}_0}. \tag{9}$$

Author Index

Lecture Notes in Computer Science

For information about Vols. 1–3269

please contact your bookseller or Springer